MOBILIZING HOSPITALITY

Mobilizing Hospitality

The Ethics of Social Relations in a Mobile World

Edited by

JENNIE GERMANN MOLZ
College of the Holy Cross, MA, USA

and

SARAH GIBSON
University of Surrey, UK

LONDON AND NEW YORK

First published 2007 by Ashgate Publishing

Published 2016 by Routledge
2 Park Square, Milton Park, Abingdon, Oxfordshire OX14 4RN
711 Third Avenue, New York, NY 10017, USA

First issued in paperback 2016

Routledge is an imprint of the Taylor & Francis Group, an informa business

British Library Cataloguing in Publication Data
Mobilizing Hospitality: the ethics of social relations in
a mobile world
1. Hospitality 2. Courtesy 3. Social ethics 4. Strangers
I. Molz, Jennie Germann, 1969– II. Gibson, Sarah, 1977–
177.1

Library of Congress Cataloging-in-Publication Data
Mobilizing hospitality : the ethics of social relations in a mobile world / edited by Jennie Germann Molz and Sarah Gibson.
p. cm.
Includes bibliographical references and index.
ISBN-13: 978-0-7546-7015-5
1. Hospitality. 2. Social ethics. 3. Emigration and immigration. I. Molz, Jennie Germann, 1969- II. Gibson, Sarah, 1977-
BJ2021.M58 2007
177'.1--dc22

2007014529

ISBN 13: 978-1-138-25015-4 (pbk)
ISBN 13: 978-0-7546-7015-5 (hbk)

Contents

Notes on Contributors *vii*
Acknowledgements *xi*

1 Introduction: Mobilizing and Mooring Hospitality 1
Jennie Germann Molz and Sarah Gibson

Seville to Hackney: A Photographic Journey 27
Elly Clarke

2 Moments of Hospitality 29
David Bell

3 Hospitality and Migrant Memory in Maxwell Street, Chicago 47
Tim Cresswell

4 Cosmopolitans on the Couch: Mobile Hospitality and the Internet 65
Jennie Germann Molz

5 Sensing and Performing Hospitalities and Socialities of Tourist Places: Eating and Drinking Out in Harrogate and Whitehaven 83
Viv Cuthill

6 Hospitality, Kinesthesis and Health: Swedish Spas and the Market for Well-Being 103
Tom O'Dell

7 Resident Hosts and Mobile Strangers: Temporary Exchanges within the Topography of the Commercial Home 121
Paul Lynch, Maria Laura Di Domenico and Majella Sweeney

8 Hospitality in Flames: Queer Immigrants and Melancholic Be/longing 145
Adi Kuntsman

9 'Abusing Our Hospitality': Inhospitableness and the Politics of Deterrence 159
Sarah Gibson

10 Hospitality and the Limitations of the National 177
Karima Laachir

11 Figures of Oriental Hospitality: Nomads and Sybarites 193
Judith Still

Index *211*

Notes on Contributors

David Bell Senior Lecturer in Critical Human Geography and leader of the Urban Cultures & Consumption research cluster in the School of Geography at the University of Leeds, UK. His research interests include spaces of hospitality, cultural policy, cybercultures and urban and rural cultures. Recent and forthcoming books include *Small Cities*, *Cyberculture Theorists*, a 2nd edition of *The Cybercultures Reader*, and *Urban Erotics*, all published by Routledge.

Elly Clarke Clarke has a Masters in Fine Art from Central Saint Martins and a First Class degree in History of Art from Leeds. Since 2003 she has worked on several projects in collaboration with the Centre for Mobilities Research at Lancaster University. She has exhibited in the UK and abroad.

Tim Cresswell Professor of Human Geography, Royal Holloway, University of London. Author of *On the Move*: *Mobility in the Modern Western World* (Routledge, 2006), *Place: A Short Introduction* (Blackwell, 2004), *The Tramp in America* (Reaktion, 2001), and *In Place/Out of Place*: *Geography, Ideology and Transgression* (University of Minnesota Press, 1996). He is currently working on the intersections of place and mobility at four sites in Chicago.

Viv Cuthill Tourism researcher and consultant. Viv recently completed her PhD in the Centre for Mobilities Research, Lancaster University. She researches transformations of tourist places and service cultures and has authored a chapter in the edited collection *Tourism Mobilities* (Routledge, 2004).

Maria Laura Di Domenico Research Fellow in Social Enterprise, Judge Business School, University of Cambridge. Her research interests include critical management studies and the sociology of organizations, social/community enterprises and social entrepreneurship, and heritage tourism. She received her PhD in 2003 from the University of Strathclyde, which explored the views, self-definitions and lifestyle orientations of proprietors of home-based hospitality enterprises in Scotland. She has published her work as journal articles, book chapters and research reports, and has presented her work at international and national conferences.

Jennie Germann Molz Assistant Professor of Sociology, College of the Holy Cross, Worcester, Massachusetts. In addition to publishing articles on the topics of travel, technology and belonging in journals such as *Citizenship Studies*, *Environment and Planning A*, and *Body & Society*, Jennie has authored chapters in several edited collections including *Emotional Geographies* (Ashgate, 2005), *Culinary Tourism*

(University of Kentucky Press, 2004) and *Tourism Mobilities* (Routledge, 2004). She recently completed a research fellowship with the Centre for Mobilities Research at Lancaster University.

Sarah Gibson Lecturer in Cultural Studies, University of Surrey. Sarah's research focuses on issues of national identity and culture, racism, and tourism. Sarah's research has been published in journals such as *Journal for Cultural Research*, *Third Text*, *Space and Culture*, and *Tourist Studies*.

Adi Kuntsman Adi has completed her PhD at the Department of Sociology at Lancaster University where she also lectures part time. Her dissertation on Russian-speaking queer immigrants in Israel/Palestine explores the relations between nationalism, ethnicity and sexuality and the ways they constitute an on-line community. Adi's research brings together theories of migration and diaspora; post-colonial and feminist theorizing of nation and nationalism; and gay and lesbian studies.

Karima Laachir Lecturer in Cultural Theory, University of Birmingham. Her research interests focus on the issues of post-war immigration and racism in Europe, with particular focus on the French case, European postcolonial diasporas and the politics of their cultural production, questions of citizenship, national belonging, cultural identity, Islam and the rise of Islamophobia in France, post-colonialism and French colonial legacy. She is also interested in 'Beur' literature, North African Francophone literature and Postcolonial studies. She is currently working on a book entitled *'Beur' Diasporic Literature*: *Rethinking Identities*, *Negotiating Belonging* (Brill Publishers).

Paul Lynch Reader in Critical Hospitality Studies, Department of Hospitality and Tourism Management, University of Strathclyde. Paul has published in journals in the fields of hospitality, tourism, service sector, leisure and entrepreneurship. His PhD involved an autobiographical sociological analysis of the guest experience of commercial home accommodation. Current research interests address commercial homes, hospitality studies and lifestyle entrepreneurship. Co-editor of *Hospitality: A Social Lens* (Elsevier, 2007).

Tom O'Dell Associate Professor in the Department of Service Management, Lund University, Campus Helsingborg. Previously, he has published *Culture Unbound: Americanization and Everyday Life in Sweden* (Nordic Academic Press, 1997). He has also edited several volumes on tourism and the experience economy including, *Nonstop*! *Turist i upplevelseindustrialismen* (Historiska Media 1999), *Upplevelsens materialitet* (Studentlitteratur, 2002), and *Experiencescapes*: *Tourism*, *Culture*, *and Economy* (Copenhagen Business School Press, 2004, together with Peter Billing).

Judith Still Chair of French and Critical Theory at the University of Nottingham. She is the author of *Justice and Difference in the Work of Rousseau* (Cambridge, 1993) and *Feminine Economies: Thinking Against the Market in the Enlightenment* *and the Late Twentieth Century* (Manchester, 1997). She is the editor of *Men's Bodies* (Edinburgh, 2003) and also co-editor (with M. Worton) of *Intertextuality* (Manchester, 1990) and *Textuality and Sexuality* (Manchester, 1993) and (with D. Knight) of *Women and Representation* (Nottingham, 1995) and (with S. Ribeiro de Oliveira) of *Brazilian Feminisms* (Nottingham, 1999). She is also an editor of *Paragraph*. She is currently holding a Leverhulme Major Research Fellowship (2004–07) to pursue research on theories and representations of hospitality in the eighteenth century and the contemporary period.

Majella Sweeney PhD researcher. Majella is a doctoral candidate in the Department of Business and Enterprise at Queen Margaret University, Edinburgh. She has research interests in commercial home enterprises, small business and hospitality. Her thesis investigates the host's relationship with the commercial home.

Acknowledgements

This book was inspired by the intensive discussions about hospitality and hostility, ethics and politics, tourism and migration, and mobility and stillness that occurred during the 'Mobilizing Hospitality: The Ethics of Social Relations in a Mobile World' workshop that took place at Lancaster University in September 2005. We wish to express our gratitude to the Centre for Mobilities Research and The Institute for Advanced Studies at Lancaster University for providing the funding, facilities, and administrative support that made this workshop possible. We are especially grateful to Pennie Drinkall for her help organizing the event. We would also like to thank the speakers from the workshop, most of whom have contributed chapters to this book, and the participants for their engaging, challenging and uplifting discussions that stimulated us to think through the mobility of hospitality within an interdisciplinary context, but also for their belief in the hope and promise of hospitality. Special thanks are due to John Urry for encouraging us to develop our work on this topic; Robert Crawshaw for graciously stepping in to help us synthesize the round table discussion; Soile Veijola whose mobile hospitality made the event challenging and fun; and Ghassan Hage for propelling us forward with his ideas, his questions, and his hopefulness.

Both the editors would like to extend a special thank you to Anne-Marie Fortier, Mimi Sheller, and John Urry for their generous intellectual hospitality and to Elly Clarke for bringing her artistic perspective and a vivid visual dimension to this project. Jennie would like to thank her family, and especially Martin and Elliot Molz, for making 'home-in-mobility' possible. Her participation in this project was made possible through a Postdoctoral Fellowship in the Centre for Mobilities Research funded by the Economic and Social Research Council (Award PTA-026-27-0623). Sarah would like to thank Dawn Llewellyn for her never-ending willingness to discuss 'hospitality', and Pat Gibson, Laura, Dan and Elizabeth Kellaway, and Sven Steenfeldt-Kristensen for their continual support and encouragement.

JGM and SEG
February 2007

Untitled **(Vienna, Austria, 2004) by Elly Clarke**

Chapter 1

Introduction: Mobilizing and Mooring Hospitality

Jennie Germann Molz and Sarah Gibson

Hospitality is a profoundly evocative concept that reverberates with cultural, political and ethical undertones. It conjures up a jumbled collage of images and senses drawn from ancient mythology, cultural traditions, scriptural references, tourism metaphors, regional stereotypes, national narratives, and government policies. Hospitality reveals its complex nature in a range of places, moments, objects and fantasies, from the material gestures of a warm smile, laden table or cosy bed, to the moral tales of Philemon or the Good Samaritan, to the iconic symbols of an open door or of the Statue of Liberty welcoming the world's tired, poor, huddled and homeless masses. At the same time, the concept of hospitality embodies its own impossibility, calling to mind images of exclusion, closure and violence: walled borders, gated communities, asylum detention centres, and race riots.

Hospitality is a phenomenon that, even in its failure, evokes the ancient and persistent question: how should we welcome the stranger, the sojourner, the traveller, the other? Where might hospitable encounters occur, and what kinds of spaces does hospitality produce? Who is able to perform the welcoming host, and who can be admitted as a guest? And in extending hospitality to the other, how should we define our individual, communal, or national self? It is toward these questions that the contributors to this book direct their investigations of a variety of phenomena, practices, places, and histories of intersecting hospitalities and mobilities. By invoking the concept of hospitality, the chapters presented here aim to reflect critically upon the ethical implications, including the limits and the possibilities, of social relations between people in an increasingly mobile and globalized world.

As individuals and groups of people now travel across the world at ever-escalating distances, scales, and speeds, the contemporary global condition is perhaps best understood through metaphors of scapes, flux, flow, mobility and liquidity (Appadurai, 1996; Castells, 1996; Bauman, 2000; Urry, 2000). Social relations are increasingly produced through mobile networks of environmental, cultural, social and economic interdependencies that transcend territorially bounded societies or nation-states (Urry, 2000; Hannam, Sheller and Urry, 2006). People and places across the globe are now bound together through complex and fluid connections that emerge around the transnational flows of commodities and capital, images and information, ethnicity and culture, crime, disease, waste and pollution. And, of course, people. New patterns of migration, diaspora, and transnational labour, along with the exponential growth of business travel and global tourism, now account for unprecedented levels of international mobility.

Even individuals who are not physically on the move may find themselves imaginatively or virtually mobilized, especially as the Internet and new mobile communication technologies bring geographically dispersed social networks together in new ways (Bauman, 1998; Tomlinson, 1999; Morley, 2000; Urry, 2000). At the same time, neighbourhoods, communities and nations are 'internally globalized' and 'cosmopolitanized' by complex global circulations of commodities and cultures that accompany the flows of tourism and migration (Massey, 1994; Beck, 2000). For those who travel as well as for those who ostensibly stay home, social life is increasingly comprised of 'strange encounters' (Ahmed, 2000). These new intersections and proximities bring the provocative dilemma of hospitality – how do we welcome the stranger? – urgently back to centre stage, reframing it against the contemporary concerns of a mobile world.

The plethora of different journeys in today's mobile world has thus led to a diversity of hospitalities. By examining both literal and metaphorical examples of hospitality, this book introduces questions of context, historicity, temporality, space, mobility, and social relations in order to complicate the meaning of hospitality. The concept of hospitality has been applied in several disciplinary settings and across a wide range of phenomena. Hospitality has long been a focus of anthropological enquiry, with ethnographic accounts framing hospitality as a way of negotiating kinship, friendship and hostility (Selwyn, 2000). More recently, the metaphors and paradigms of hospitality have emerged in other fields as well, from historical accounts of the shifting social and cultural meanings of hospitality (Heal, 1990; Browner, 2003), to commercial forms of hospitality, such as those provided by the travel and tourism industry (Smith, 1989; Lashley and Morrison, 2000; Lynch, 2005; Lashley, Lynch and Morrison, 2007), to less explicit forms of hospitality extended by the nation to migrants, refugees or asylum seekers (Ahmed, 2000; Rosello, 2001; Pugliese, 2002; Schlunke, 2002; Gibson, 2003, 2006; Yegenoglu, 2003; Chan, 2005; Metselaar, 2005; Savic, 2005; Kelly, 2006; Worth, 2006). Along these lines, critical accounts of hospitality in the context of business, tourism and migration have also sought to highlight the contingent notion of hostility within hospitality (Gibson, 2003). The concept of hospitality, especially in its Kantian articulation, has been revived to address human rights and cosmopolitical formulations of a universal law of hospitality (Derrida, 1999, 2000: 2001b; Honig, 1999; Dikeç, 2002; Venn, 2002; Vertovec and Cohen, 2002; Amin, 2004; Benhabib, 2004, 2006), and emerging forms of online social relations and cybernetic encounters have been examined in terms of technological forms of hospitality and belonging online (Aristarkhova, 1999, 2000). Clearly, the concept of hospitality poses practical and theoretical questions that span disciplinary boundaries.

A key aim of this collection of essays is to enact a form of 'intellectual hospitality' (Kaufman, 2001; Bennett, 2003) by considering how the deployment of the concept of hospitality in one disciplinary context may provide insights in another. As Friese argues 'what is at stake is not only the thinking *of* hospitality, but thinking *as* hospitality' (2004, 74; and see Still, 2004). In the able hands of various scholars, the cultural, commercial, philosophical, political, ethical and social dimensions of hospitality have been subjected to rigorous debate. Yet, perhaps because of its wide appeal across disciplines, the concept of hospitality has eluded any attempt to delineate it as a unified theoretical paradigm or ontological framework. Nor is that our intention here. However, we do aim to coordinate some of these interdisciplinary approaches to hospitality through the perspective of mobility.

Hospitality is not just a metaphor for reflecting on encounters with the stranger (Rosello, 2001), but serves more broadly as a central concept for the emergent paradigm of 'mobilities' (Urry, 2000). By focusing on the complex intersection between hospitality and mobility, we hope to open up a space for imagining humane and ethical answers to the pressing question of how to welcome the stranger in today's mobile world. Hospitality is a structure that regulates, negotiates, and celebrates the social relations between inside and outside, home and away, private and public, self and other (Still, 2006: 704). We seek to engage critically with the way such social relations are structured through literal or metaphorical gestures of hospitality that transgress and reiterate boundaries, that fix and mobilize identities, and that negotiate the distinction between the self and the other. Thus, this book is about the 'politics of mobility' (Cresswell, 2001) and the 'ethics of hospitality' (Derrida, 1999).

Implicit in most definitions of hospitality are the movements of tourists and visitors (those mobile others who come and go) as well as the movements of migrants, asylum seekers, and refugees (those mobile others who come and stay). Although mobility underpins any discussion of hospitality, none of the erstwhile contributions on hospitality have explicitly brought a mobilities focus to bear on hospitality, nor has the emerging scholarship on mobilities studies included a sustained focus on the political, philosophical and ethical aspects of hospitality. The notion of hospitality, as Hent de Vries (2001) argues, has 'immense relevance [...] for the most urgent questions dominating contemporary political debates' on immigration, globalization, multiculturalism, and citizenship (178). In this book, we want to make explicit what is often implicit about hospitality – its predication on mobility – and to highlight the fluidity of the practices and categories of hospitality. In order to do this, the book inserts itself at some busy intersections: between politics and ethics, between tourism and migration, between travelling and dwelling, and between mobility and immobility.

The Politics of Mobility and the Ethics of Hospitality

To be sure, the question of hospitality is not new; indeed, it is one of human civilization's most ancient themes. Nor is the global movement of people in itself a new phenomenon. For centuries, the movements of traders, travellers, pilgrims, tourists, migrants, explorers, nomads, colonialists and warriors have drawn individuals, communities and nations into contact along varying degrees of friendship and violence. Much scholarly attention has been paid to the question of how the social relations that occur in these 'contact zones' (Pratt, 1992) should be politically or ethically negotiated.

In recent articulations of these debates, Immanuel Kant's *Toward Perpetual Peace* (1996 [1975]) has been of particular importance in teasing out the complex civic and moral implications entwined in questions of mobility, citizenship and human rights. In this slim but influential treatise, Kant observes that: 'by virtue of the right of possession in common of the earth's surface on which, as a sphere, [humans] cannot disperse infinitely but must finally put up with being near one another' (1996 [1795], 329). In other words, because humans inhabit a geographically limited planet, it is our natural destiny to come into contact with one another. For Kant, this 'natural law' of shared residence on the earth's surface assumes a 'cosmopolitan right' to travel and encounter each other under various auspices. This right that is conditioned by the law of 'universal hospitality', which ensures the rights and duties associated with the movement of foreigners around the world: the right to travel and be received in other lands without hostility; and a duty to not use one's travels as a means of exploitation or oppression (see Waldron, 2006: 90). Written more than 200 years ago, Kant's reflections on global civil society sought to institutionalize a special relationship between mobility and hospitality as the underlying tenet of cosmopolitan interaction between people and nations. We maintain that this intersection between mobility and hospitality is just as relevant today, if not more so, in framing the political and ethical parameters of social interaction, moral duties and state obligations in a world of strange encounters. And we are not alone. Scholars across disciplines have revived Kant's notions of universal hospitality and cosmopolitan right to address contemporary concerns, especially around issues of migration, asylum and citizenship (see Nussbaum, 1994; Derrida, 1999; Bauman, 2002; Dikeç, 2002; Benhabib, 2006; Waldron, 2006). While the current debates owe much to Kant's meditations on hospitality, many scholars are critical of the juridical conditions and necessary limitations that Kant imposes on the concept of hospitality. One of the more powerful critiques has been offered by Jacques Derrida in his various writings on cosmopolitanism and hospitality.

Because several of the chapters in this collection engage directly with Derrida's work on hospitality, we want to take a moment here to outline Derrida's critique of Kant's universal hospitality and to reflect on Derrida's contribution to our understanding of hospitality as a framework for thinking about the ethics of social relations in a mobile world. Derrida explains that because Kant's notion of hospitality relies on conditions of reciprocity, duties and obligations between people and nation-states it delimits rather than opens up borders and possibilities. Derrida admonishes that Kant's hospitality is 'only juridical and political: it grants only the right of temporary sojourn and not the right of residence; it concerns only the citizens of States' (Derrida, 1999: 87).

In contrast, Derrida draws a distinction 'between an *ethics* of hospitality (an *ethics as* hospitality) and a *law* or a *politics* of hospitality' (Derrida, 1999: 19), seeing Kant's formulation of hospitality as a politics of conditional hospitality as opposed to an ethics of infinite, unconditional and absolute hospitality (Gibson, 2003). The laws of hospitality place a series of conditions upon the welcoming of others, but *the* law of hospitality – hospitality as an ethics – 'tells us or invites us, or gives us the order or injunction to welcome *anyone*, any other one, without checking at the border' (Derrida and Düttmann, 1997: 8).

What Derrida encourages us to think about is a hospitality that is infinite, absolute and completely open – a welcoming of the other and regardless of who that other is, regardless of the potential dangers and risks involved. An ethics of hospitality entails opening one's borders or doors to *anyone*, acting beyond our own self-interest. It is not an easy thing to imagine, and indeed Derrida is fully aware of this difficulty. As Gibson observes:

> Absolute hospitality is impossible as it undermines the very condition of a nation or state, which is constituted through the erection of frontiers and borders. Absolute hospitality requires the "generosity" of the state even as the ethical notion of absolute hospitality goes beyond any frontier or border of the state (2003: 374–375).

Absolute hospitality is impossible for the nation-state, and equally aporetic in the case of interpersonal exchanges of hospitality, for in welcoming the foreigner unconditionally, the host must relinquish the mastery of his or her home which is the condition of being able to offer hospitality in the first place. In other words, absolute hospitality requires us to go beyond, even beyond the very conditions that enable a state or a person to offer hospitality at all.

Derrida is concerned with the difficulty in thinking through these two supplementary meanings of hospitality as an ethics and as a politics.

> If the two meanings of hospitality remain mutually irreducible, it is always in the name of pure and hyberbolic hospitality that it is necessary, in order to render it as effective as possible, to invent the best arrangements [dispositions], the least bad conditions, the most just legislation. This is necessary to avoid the perverse effects of an unlimited hospitality whose risks I tried to define. This is the double law of hospitality: to calculate the risks, yes, but without closing the door on the incalculable, that is, on the future and the foreigner (Derrida and Düttmann, 2005: 6).

His concern is not to reconcile the politics of hospitality with an ethics of hospitality, but rather to extend a provocative challenge that speaks to the politics of self-other relations and draws out a model for living with difference.

As critics working especially in the area of migration and multiculturalism remind us, our official and informal policies toward welcoming the other for the most part fall far short of Derrida's ideal of absolute hospitality (see Gibson in this volume). While we might find in political and popular rhetoric gestures toward multiculturalist tolerance and metaphors of generous hospitality surrounding the reception of migrants, these discourses often serve to reiterate a specific power relation between the self and the other. As Yegenoglu (2003) notes, 'far from laying the grounds for an interruption of sovereign identity of the self, multiculturalist respect and tolerance implies the conditional welcoming of the guest within the prescribed limits of the law and hence implies a reassertion of mastery over the national space' (16). In other words, hospitality tends to reassert the identity and belonging-ness of the host against the movement, shifting, unstable, un-belonging-ness of the guest. But in Derrida's deconstruction of hospitality, the binary opposition between host and guest unravels:

> The *hôte* who receives (the host), the one who welcomes the invited or received *hôte* (the guest), the welcoming *hôte* who considers himself the owner of the place, is in truth a *hôte* received in his own home. He receives the hospitality that he offers in his own home; he

> receives it from his own home – which, in the end, does not belong to him. The hôte as host is a guest (Derrida, 1999: 41).

Like Derrida, we want to destabilize hospitality as a paradigm and 'host' and 'guest' as distinct categories, by 'mobilizing hospitality' – by opening it up and by questioning its closures, by examining the nuanced fluidity of categories such as host and guest, and by disassociating stasis with hosts/homes and movement with guests/travel. We take as our starting point the mobilities of tourism and migration, which are generating new patterns of circulation, intersection and proximity between strangers. The chapters in this book bring debates around voluntary and obligatory mobilities into conversation by examining the politics of travelling and staying still and by interrogating the ethical responses to mobile others who are more or less invited, more or less welcome.

Tourism and Migration: Circulating Discourses of Hospitality

In distinct yet related ways, the academic, political and popular discourses surrounding tourism and migration have invoked the metaphors of hospitality – along with terms such as host, guest, welcome, refuge, invitation and home – to categorize people, to erect, police and transcend symbolic and physical boundaries, and to authorize or condemn certain terms of social interaction. Much of the current literature that applies a hospitality framework to studies of tourism or migration mobilities has tended to reproduce fairly rigid social categories that associate mobility with the guest, tourist or migrant, and immobility with the host, local population, home or nation. The chapters in this book seek to destabilize these associations by demonstrating that these categories are produced through intersecting circulations of mobility and rest, people and places, objects and memories (see chapters by Bell, Lynch, Di Domenico and Sweeney, and Cresswell in this volume). Here, we highlight some of the specific overlaps and disjunctures between tourism and migration discourses in negotiating the political and ethical implications of practices of welcoming, ways of belonging, and forms of citizenship while on the move.

Hospitality in Tourism Discourse

Since the publication in the late 1970s of Valene Smith's influential collection *Hosts and Guests*, hospitality has been one of the most pervasive metaphors within tourism studies, referring in one sense to the commercial project of the tourist industry (such as hotels, catering, and tour operation) and in another sense to the social interactions between local people and tourists – that is, hosts and guests. Smith's influential collection helped to reframe the focus of tourism studies away from an emphasis on tourists and toward a critical concern with the unequal social relations within tourism and the impact of tourism on local populations and environments (McNaughton, 2006). Over the years, many scholars have sought to revise and refine the deceptively simple paradigm, while others have rejected the usefulness of the hospitality metaphor altogether.

Empirical studies consistently indicate that the binary oppositions between the categories of host and guest rarely hold up in the field (McNaughton, 2006). For example, cases of diasporic migrants revisiting their 'homeland' as tourists (Duval, 2003, 2004); second-home owners travelling or retiring to their holiday homes abroad (O'Reilly, 2003; Hall and Mueller, 2004); migratory labourers working in the hospitality industry (Choi, Woods and Murrmann, 2000); travellers employed on working holiday programs (Clarke, 2004; Heuman, 2005); the seasonal influx of handicraft and souvenir merchants to tourist areas (McNaughton, 2006); or multinational tourist developments funded by overseas interests (Selwyn, 1996; Tribe, 2005) all call into question who is the host and who is the guest. Many researchers have challenged the binary opposition between host and guest by refining these categories in more pluralistic and heterogeneous terms, or by inserting a continuum of social actors, including observers, brokers and mediators, between the two poles of host and guest (see Crick, 1985; Selwyn, 1996; Cheong and Miller, 2000; McNaughton, 2006).

More useful are critiques that frame host and guest as fluid, contested social roles that people move into, out of, and in between as they negotiate extensive overlapping mobilities and social memberships (Sherlock, 2001; Duval, 2003). For example, O'Reilly (2003) shifts the frame of reference by asking '*When* is a tourist?' suggesting that social categories such as tourist and migrant, or indeed host and guest, can be seen as temporally-constrained social performances rather than as strictly-bounded identity categories (see Hall and Jenkins, 1995; Sherlock, 2001; Duval, 2003; McNaughton, 2006). For our purposes, O'Reilly's question may also be helpfully rephrased as '*Where* is a host or guest?' to remind us that hospitality produces not only certain social categories (such as host and guest), but also certain spatial patterns of social interaction on the move (see in particular chapters by Bell, Cresswell, and Cuthill in this volume).

Some critics find the host-guest paradigm altogether irrelevant to contemporary forms of tourism. For example, Aramberri (2001) urges the host to 'get lost', arguing that a hospitality framework fails to adequately account for the commercial organization of modern mass tourism and the fact that, for better or for worse, most interactions between tourists and local people come down to some form of monetary exchange. In his estimation, the terms 'service providers' and 'customers' more accurately capture these social interactions (Aramberri, 2001: 746). He goes on to itemize the various ways in which mass tourism, even when posing as sustainable development, transforms touristic encounters into commercial exploits that have no relation to 'the old convenant' of hospitality (746). And yet, in many circles, hospitality has become synonymous with the tourism industry, referring precisely to the commercial provision of tourist services (see Lashley and Morrison, 2000; Lashley, Lynch and Morrison, 2007; and see Cuthill and O'Dell in this volume). Given the expanding realm of activities that seem to count as hospitality within tourism research and the tourism industry, we suggest that instead of rejecting the hospitality paradigm altogether, we should consider the way the concept of hospitality continues to be made meaningful within academic discourse and everyday practices. For example, Cuthill (Chapter 5) discusses the production of hospitality in restaurants and bars in two English cities, while O'Dell (Chapter 6) analyses spas in Sweden as sites of hospitality.

More importantly, we need to pay attention to the way the concept of hospitality brings meaning to certain social arrangements between strangers. For example, instead of rejecting commercialized interactions as *not* hospitality, the contributors to this book examine the way the ethics of social interactions in commercial settings are contested through a discourse of hospitality. For example, in her analysis of online hospitality clubs, Germann Molz (Chapter 4) finds that travellers evoke ideals of hospitality precisely to contest commercialized uses of the Internet and commercialized host-guest relations. Meanwhile, Lynch, Di Domenico and Sweeney (Chapter 7) tease out the complex overlaps between for-profit and private hospitality in the context of the 'commercial home'. Commercial venues such as restaurants, spas and bars do not elide hospitality, but rather become the setting for complex negotiations of multiple hospitalities and embodied encounters between strangers (see chapters by Cuthill, O'Dell, and Kuntsman in this volume).

Instead of jettisoning the paradigm, we should take this opportunity to ask who gets to be a guest, and under what conditions? Who gets to be a host, and under what conditions? Who gets to move between these categories? How do these categories authorize some people's right to travel and to be welcomed, while delegitimizing other claims to mobility and belonging? As McNaughton rightly notes, 'host and guest are not innocent terms. They reverberate with a sense of welcome and with particular understandings of hospitality that reveal more about the archaeology of the discourse than they do about the nature of contemporary tourism' (2006, 648).

The host-guest paradigm has certainly provided tourism studies with a powerful framework for interrogating social relations within the domain of leisure mobilities, but it has recently been evoked in the very different context of national discourses around migration and asylum seeking. This is certainly not to say that the boundaries between tourism and migration are patently distinct; indeed, recent studies have demonstrated the way tourism and migration mobilities overlap and inform each other (see Feng and Page, 2000; Hall and Williams, 2002; Duval, 2003; O'Reilly, 2003; Coles and Timothy, 2004). Attending to the relations between 'migration, return migration, tourism, transnationalism and diaspora' helps us to focus on the uneven mobilities of 'obligatory as well as voluntary forms of travel' (Hannam, Sheller and Urry, 2006: 10) and to consider the way such unevenness is reproduced through discourses of hospitality that determine which strangers are (or are not) invited and welcome.

Migration Discourse

The metaphor of hospitality structures contemporary debates on nationalism, migration, multiculturalism, and asylum. Who feels at home within the nation? Who is excluded or fails to feel at home in the nation? Is a host necessarily a citizen of the host nation-state? Why are immigrants, refugees, and asylum seekers imagined as guests of the host nation-state? These are important questions for understanding the metaphors of hospitality and the home in contemporary debates on national identity and citizenship (see Kelly, 2006). Hospitality is intimately connected to nationalism, where crossing the border into the nation (whether as an immigrant or as a tourist) is dependent upon national definitions of what counts as hospitality, and the figure towards whom hospitality is offered and received (Rosello, 2001, viii). In the context of debates on nationalism and immigration, discourses of hospitality work to blur 'the distinction between a discourse of rights and a discourse of generosity, the language of social contracts and the language of excess and gift-giving' (Rosello (2001: 9). In these debates, the Kantian cosmopolitan right to 'universal hospitality' is in tension with the sovereignty of the nation-state (see Benhabib, 2004, 2005).

In studies of migration, multiculturalism and postcolonialism, the metaphor of hospitality is frequently invoked (Ahmed, 2000, 2004; Rosello, 2001; Hage, 2002, 2003; Chan, 2005; Still, 2006). But this metaphor of hospitality is a dead metaphor (Rosello, 2001: 3) since such studies employ the metaphor of 'hospitality' precisely to reveal the hostility present within such policies of managing diversity within the 'host nation'. In constructing 'the immigrant as guest' (Rosello, 2001), the host nation excludes the immigrant from feeling at home in the nation. This opposition between host/guest, native/stranger maintains the line between power/powerlessness, ownership/dispossession, stability/nomadism (Rosello, 2001: 18). Such a rhetoric of hospitality is ideological as it enables 'some people to have *fantasies of control*' (Hage, 2002: 165; see Gibson, this volume) in the power to host and welcome.

Similarly multicultural national imaginaries which often employ the metaphor of hospitality are revealed to be, in fact, 'not very hospitable' (Ahmed, 2000: 190) as they continue to position 'the natives' as hosts who decide which guests/strangers will or will not be welcomed. Discourses of multiculturalism involve the contradictory processes of 'incorporation and expulsion' (Ahmed, 2000: 97) or an 'inclusive exclusion' (Laachir this volume). The guests/strangers in such a narrative of multiculturalism are consequently placed under a 'debt of hospitality' (Chan, 2005: 21) to the host nation. Such uses of the metaphor of hospitality in studies of migration and multiculturalism similarly ignore the historical social relations of colonialism, which involved the transformation of guests into hosts (Ahmed, 2000: 190). Whether the host nation welcomes, expels, or deters the stranger these responses to the other are all premised on the same power relation. It is the native who is empowered to feel at home and to assume the position of the host. If the immigrant is imagined as 'the guest,' the 'host nation' maintains its historical position of power and privilege in determining who is or is not welcome to enter the country, but also under what conditions of entry. Hospitality, however, is not simply a question of crossing (or not) the border. The question today, Bauman argues, is how to live with strangers daily and permanently (1997: 55).

The host nation, despite explicit evidence to its contrary, often imagines itself narcissistically as being hospitable. Derrida's distinction between a limited, conditional hospitality and an infinite, unconditional hospitality has been critically engaged with to puncture these narcissistic myths nations use to construct the current so-called problem of asylum (on Britain see Ahmed, 2004, and Gibson, 2003, 2006, and in this volume; on the Netherlands see Metselaar, 2005; on France see Rosello, 2001, and Still, 2004; on Australia see Kelly, 2006, Pugliese, 2002, and Schlunke, 2002; and on New Zealand see Worth, 2006). In such studies, the figure of the asylum seeker is constructed as 'the uninvited' (Harding, 2000), where the nation-state imagines itself to be a 'reluctant host' (Joly and Cohen, 1989) who is unwilling to generously offer hospitality to such unwelcome and parasitical guests. The tension between the human right to asylum (which is ratified in international agreements) is often in contrast to the right of the nation-state to maintain control over its borders.

While the metaphor of hospitality in discourses of nationalism and immigration has empowered the native to assume the powerful position of the host, it is precisely this metaphor that needs to be deconstructed in order to conceive new ways of figuring the social relations between citizens, immigrants, refugees, asylum seekers and nation-states. The metaphor of hospitality needs to be deconstructed in order to interrogate the different contexts in which it is deployed as a means of legitimating the power of some while disavowing the rights of others. If the immigrant is imagined as a guest (Rosello, 2001), the figure of the immigrant is conceived either negatively in anti-immigration discourses as a parasite or positively in discourses of multiculturalism as a grateful guest. While the host-guest paradigm has been useful in theorizing social relations between strangers within studies of nationalism, immigration, and multiculturalism, rather than imagining the immigrant through the binary opposition of host/guest it is important to re-conceive the social relations that characterize the relationships between host and guest, citizen and immigrant. Hospitality is about the other questioning and interrupting the self, rather than reasserting the mastery of the self. Instead of rejecting the metaphor of hospitality, the contributors to this book take the opportunity to consider the promise of hospitality (see the chapters by Gibson, Kuntsman, Laachir, and Still in this volume) in reconfiguring social relations between strangers within studies of nationalism, immigration, and multiculturalism.

A key point of intersection between the discourses we have just described is the way the concept of home is evoked in the ethics and politics of welcoming the other. National discourses of hospitality frame the nation-state as a 'home' that is open to (certain) foreigners, but whose borders must be protected; while in tourism, the notion of hospitality suggests a range of possible homes, including the cities and local places tourists visit, the homes of friends and family members who host travellers, or the hotel or resort that serves as the tourist's 'home-away-from-home'. Tourism and migration mobilities both imply a movement away from home, but also toward a new (permanent or temporary) home. For example, migration studies often 'foreground acts of "homing" and "re-grounding" which point towards the complex interrelation between travel and dwelling' (Hannam, Sheller and Urry, 2006: 10; and see Hage 1997 on 'migrant home-building' and Brah 1996 on diasporic 'homing desires'). The chapters in this book suggest that as much as hospitality is associated with mobility, it is equally concerned with stasis and rest (a place to eat, sleep, or recuperate). Indeed, hospitality occurs precisely at this intersection between travel and dwelling. To host or to be hosted are both forms of travelling-in-dwelling and dwelling-in-travelling where the mobilities of guests, travellers and foreigners intersect with hosts and homes.

Travelling and Dwelling

Home is often idealized as a 'space of hospitality' (Dikeç, 2002; Friese, 2004) that offers the traveller respite from the labour of mobility. Where is home? What does it mean to be at home? Who feels at home and who fails to feel at home? Who can be a mobile host away from home, or a guest at home? Edith Wyschogrod has argued that 'for there to be hospitality there must be a home' (2003, 36). Home is clearly central to formulations of hospitality, however, what constitutes home depends very much on the way hospitality is imagined, performed, offered, or denied.

If hospitality, as we suggest, poses the question of how to welcome the stranger, how to make the stranger feel 'at home,' this question assumes the figure of the stranger (who receives hospitality) as well as the figure of the host or hostess (who is able to offer hospitality), in addition to the place of hospitality. In his influential essay on 'The Stranger', Georg Simmel argues that the stranger is a 'purely mobile person' (1971, 148), who is able to move across borders, and who embodies both relations of proximity and distance within the home. In other words, the mobile stranger troubles assumed notions of home as bounded or static. The concept of home is, as David Morley argues 'the uninterrogated anchor or alter ego of all this hyper-mobility' (2000: 3). However, to focus purely on mobility would be to ignore the complex abstractions regarding the imagining of the stranger, in defining who is welcomed or expelled (Ahmed, 2000) or deterred (Gibson in this volume) from crossing the border of the nation-state or the threshold of the home. Hospitality can only be offered to the figure of the stranger, but as Sara Ahmed has so persuasively argued there are 'substantive differences between ways of being displaced from "home"' (2000: 6).

Discourses of hospitality are predicated upon home, but they also produce home – and who gets to be at home – in particular ways. For example, social groups who negotiate community and belonging on the Internet may consider a website, discussion forum or homepage to be a kind of online 'home' (see Kuntsman and Germann Molz in this volume). The hospitable city may offer to make some visitors feel at home (Bell, 2007, and in this volume), while denying this homeliness to others (see Cresswell in this volume). Or the nation itself may become a 'home' (Kelly, 2006; and see chapters by Gibson, Laachir, and Kuntsman in this volume). Just as 'host' and 'guest' are never innocent terms, neither are metaphors of home unencumbered with histories of power, especially the sovereign power to determine which mobile strangers are welcome, and which are not.

Mobility is 'a resource to which not everyone has an equal relationship' (Massey, 1994; Ahmed, 2000; Skeggs, 2004: 29) and is often dependent upon the exclusion of those who are fixed in place (Ahmed, 2004). Similarly, emphasis must be placed here on the specific contexts of travel and dwelling, for just as not all travellers travel under the same conditions, nor are people able to be at home under the same conditions. As Clifford suggests, analysis of the tensions between dwelling and travelling must account for 'specific histories, tactics and everyday practices of dwelling and traveling' (1997, 24). Who is able to make themselves at home, and under what conditions? Who is able to offer hospitality, and how does the offer of hospitality entrench certain relations of power, ownership and sovereignty? This is the power geometry of hospitality, hospitableness and hospitable social relations (see Massey 1994 on the power geometry of mobility). Hospitality is not offered to every stranger, nor does every stranger gratefully receive the gift (or debt) of hospitality. Similarly, not everyone is able to give hospitality to the stranger, not everyone is empowered to be hospitable. It is only to those recognized, identified, familiar, welcome-able strangers who are generously given hospitality, and this gesture of hospitableness can only be made by those hosts who feel at home.

Even under the guise of tolerance and generosity, hospitality frames 'home' as a domain of power where the host polices the conditions by which the front door remains open or closed. In welcoming the stranger, the host is positioned as being at home (in control of the home) in contrast to the mobility of the stranger. As Yegenoglu argues, 'hospitality is a giving gesture. But with the hospitality as law, what this gesture in fact does is to subject the stranger/foreigner to the law of the host's home. In this way, the foreigner is allowed to enter the host's space under conditions the host has determined' (2003, 15).

At the same time, however, hospitality disturbs its own framework of power. Hospitality *requires* opening up the home of the host to the other (Derrida 2000, 25). Derrida argues that hospitality is

> part of being at home; there is no home, no cultural home, no family home without some door, some opening and some ways of welcoming guests. But in that case the hospitality is conditional, in that the Other is welcome to the extent that he adjusts to the *chez soi*, to the home, that he speaks the language or that he learns the language, that he respects the order of the house, the order of the nation state and so on and so forth (Derrida 2001a: 97–98).

Hospitality is simultaneously about 'opening, without abolishing' (Dikeç, 2002: 299) the boundaries of the host's home. Thus, the gesture of hospitality reinforces the host's position of power and privilege, maintaining the host's sovereignty and control over the limits/borders of their 'home' (Derrida 2003: 127–128). John D. Caputo explains that 'when the host says to the guest, "Make yourself at home", this is a self-limiting invitation. "Make yourself at home" means: please feel at home, act as if you were at home, but, remember, that is not true, this is not your home but mine' (1997: 111).

This model of home and hospitality assumes that the home is secure against what is foreign, strange, and unfamiliar. Home, however, is not simply the place from which the stranger departs, it is also the place of arrival and transit. If the stranger is the one who has moved away from home, this neglects that there is movement implicated in the formation of homes themselves as 'complex and contingent spaces of inhabitance' (Ahmed, 2000: 88; Ahmed et al., 2003). Similarly, homely places do not simply have to be conceived as singular and bounded (see Lynch, Di Domenico and Sweeney, this volume). Criticizing the connotations of homely places as being associated with stasis, fixity, and nostalgia (homesickness), Massey instead argues that places are 'not so much bounded areas as open and porous networks of social relations' (1994: 121). Instead, we might think of home as a mobile place that is 'implicated within complex networks by which "hosts, guests, buildings, objects and machines" are contingently brought together to produce certain performances in certain places at certain times' (Hannam, Sheller and Urry (2006: 13).

Hospitality puts the notion of home into motion, opening it up literally and figuratively to the intersecting flows and circulations of hosts, guests, buildings and objects that simultaneously challenge, reassert and perform a place *as* home. In this sense, we might think of the home through the metaphor of the ship, which is for Foucault 'the heterotopia *par excellence*' (2002, 236). The ship is 'a floating piece of space, a place without a place, that exists by itself and at the same time is given over to the infinity of the sea' (Foucault, 1986: 27). Like ships, places are what Hetherington calls 'immutable mobiles', simultaneously stabilized and mobilized around networks of agents, humans, and non-humans (Hetherington, 1997: 185–189; also see Gilroy, 1993).

Hospitality as Mooring

In December 1997, several Caribbean governments, including the British territory of the Cayman Islands, denied docking privileges to a cruise ship whose passengers consisted of about 900 gay and lesbian travellers from the United States and Europe. In her analysis of the incident, Jasbir Puar explains that Cayman officials denied docking rights to the so-called gay cruise because the passengers could not be counted on to 'uphold standards of appropriate behavior' on the island (cited in Puar, 2002: 101). At the time, a US-based gay rights organization called on the British government to intervene, and then Prime Minister Tony Blair determined that the refusal of the cruise ship constituted a breach of the International Covenant of Human Rights and should therefore be rescinded.

According to Puar, the debates and discussions surrounding the affair revealed conflicting transnational alliances within the context of postcolonial politics, human rights discourse, and the economics of commercial tourism. For example, though the local Caribbean media and many local organizations stayed quiet on the topic for fear of a backlash, those activists who did speak out framed their protests in commercial terms, arguing that alienating affluent gay travellers would hurt the Caribbean tourist economy. In the end, and despite protests from activists and the British government, the cruise ship was not allowed to dock (Puar, 2002). For Puar, this incident and the ensuing debates posed questions about global and local sexualities, global gay identity in postcolonial situations, and the contested mobilities of queer tourism.

More recently, the *MV Tampa* was controversially refused hospitality on the shores of Australia. In August 2001, a Norwegian container ship, the *MV Tampa*, rescued 438 predominantly Afghani asylum-seekers from a sinking Indonesian ferry. The Australian government refused to allow the *Tampa* to dock at the Australian territory Christmas Island, arguing that the asylum seekers were the responsibility of Norway or Indonesia. Upon entering Australian waters, the *MV Tampa* was boarded by Australian Special Forces and pushed back to international waters. The *Tampa's* captain, Arne Rinnan, referred to these rescued asylum seekers as his 'guests', while Australia argued that that it would 'not be held hostage by our own decency' (in Perera, 2002). The asylum seekers were eventually transferred to Nauru, and then to New Zealand. This refusal by Australia in allowing the *Tampa* to dock reveals complex debates on hospitality, at the national, international, and maritime level. It also revealed historical debates on Australia itself.

Perera (2002: 33) charts how the 'discourse of hospitality and care for guests' was during this period asserted most strongly by Indigenous Australians as an assertion of their ownership of the land. At the same time, the Australian crisis of hospitality predicated upon this ship was invoked in New Zealand's narcissistic narratives of their superior hospitality and compassion towards asylum seekers (Worth, 2006: 226). These gestures of willingness to offer hospitality, for Perera, link 'inside and outside, these and those bodies, our stories and their stories; make simple reciprocal gestures between guest and host, sheltered and homeless, harbour and traveller' (2002, 34). Australia's failure to offer hospitality towards these asylum seekers shores up the national imaginary of Australia through the securitization of the 'boat people' as a threat, but it is also in conflict with the maritime humanitarian regime at sea which is predicated upon solidarity among seafarers (Pugh, 2004). The sea upon which the ship moves (and is moved) invokes the metaphors of flows, waves, and liquidity (Bauman, 2000; Urry, 2000) that characterize modernity. In relationship to the ships that carry potential asylum seekers and refugees towards the host nation, these strangers are imagined metaphorically as 'waves,' 'tides' or 'floods' in posing a threat to the nation but often these bodies are themselves 'at the mercy of tides, waves, shipwreck and drowning' (Pugh, 2004: 55).

We invoke these stories of ships and harbours here in part to demonstrate the complexity of negotiating hospitality through the tangled grids of global tourism, cultural mores, commercial aspirations, and social anxieties as well as international human rights, maritime regimes at sea, state sovereignty, and land ownership. Indeed, these episodes of (failed) hospitality highlight the way the privilege to be mobile, as well as the privilege to stop and visit, is always filtered through the prism of race, class, ethnicity, gender, and sexuality. However, we also invoke these incidents in order to think about the way hospitality is always implicated in the double and contradictory processes of movement and stillness embodied by the ship.

In keeping with the story of the ship, the dock, and the denied visitors, we suggest thinking of hospitality as a form of 'mooring'. Hannam, Sheller and Urry (2006) introduce the metaphor of 'mooring' to highlight the necessary interdependence between mobility systems and immobile infrastructures or platforms that enable mobilities. They argue that 'mobilities cannot be described without attention to the necessary spatial, infrastructural and institutional moorings that configure and enable mobilities', pointing to examples such as roads, transmitters, airports, docks and factories as examples of the immobile infrastructures 'through which mobilizations of locality are performed and re-arrangements of place and scale materialized' (6). The very immobility of these vast material infrastructures make mobility systems *possible*.

Hospitality is produced through the negotiation of movement and mooring. Even a world of constant mobility must sometimes stop. We need to slow down, sit down, settle and sleep. If hospitality is predicated on mobility, it is equally predicated on immobility – those places and moments of rest and repose that refresh and rejuvenate the traveller. In mobilizing hospitality, we must also attend to the constitutive immobilities that define hospitality and to the power relations that emerge at this intersection between moving and staying still. As Hannam, Sheller and Urry (2006) argue, the interplay between mobility and moorings raises important questions of 'how to move and how to settle, what is up for grabs and what is locked in, who is able to move and who is trapped' (8). These are the questions that this book aims to address – how do we think about strangers and others in a mobile world; in a world where strangers need to sit down, sleep, or settle in the presence of one another? How can we think, or re-think, hospitality through an ethical framework appropriate for our contemporary situation of mobility?

The maritime associations of the term 'mooring' call to mind an infrastructure of ropes, cables, docks, and anchors that secure ships in a safe harbour. A moored boat is anchored at dock or in a bay, safe from the heavy winds and waves of the open sea. Yet, the mooring itself must also be flexible, giving the boat some leeway to shift and move within the water. We must think about encountering the stranger as a complex negotiation between moving and staying still – of allowing mobile others a chance to stop, rest and breathe without locking them or ourselves figuratively or literally in place (see Laachir, this volume).

The metaphor of mooring moves us in this direction, suggesting as it does the notion of safe harbour, but also the possibility of (re)launching our journey. For Ghassan Hage (2002, 2003) hospitality is intertwined with hope:

> [W]hat we are talking about when it comes to discussing hospitality towards asylum seekers, or compensation for the colonised indigenous people of the world, or compassion towards the chronically unemployed [is]: *the availability, the circulation and the exchange of hope*. Compassion, hospitality and the recognition of oppression are all about giving hope to marginalised people (2003: 9).

Thus, hospitality is not just about the gift of repose, but also about the gift of hope. Making the guest feel at home is not just about seeing to his or her physical comfort or embodied needs (though these are certainly important); it is also about instilling the guest with a feeling of hope and a sense of being 'propelled' forward (Hage, 2005). As Hage has eloquently argued, hospitality provides not only a place to be safely still, but also the hope of moving:

> For what is security if it isn't the capacity to move confidently? And what is 'home' if not the ground that allows such a confident form of mobility […]. A home has to be both closed enough to offer shelter and open enough to allow for this capacity to perceive what the world has to offer and to provide us with enough energy to go and seek it (2003: 28.)

In other words, hospitality mobilizes the guest. Hospitality, home and hope are all intricately inscribed upon one another as the gift of staying still *and* moving forward.

Welcome to the Book

Taking our cues from the way scholars such as Ghassan Hage, Jacques Derrida, Sara Ahmed, Mireille Rosello and others have reflected on the ethical dimensions of hospitality, we hope to deconstruct the notion of hospitality in order to reassemble it in terms of movement, possibility and hope. We want to illuminate the paradox of hospitality – that it evokes *both* home *and* movement – in order to destabilize the way power relations are congealed within the paradigmatic discourses in tourism and migration studies. We hope to figuratively mobilize the concept of hospitality across disciplines and to de-couple associations of the host with home, territory, stability, and ownership on one side, and of the guest with mobility, estrangement, and un-belonging on the other. In casting a critical eye over this intersection between mobility and hospitality, we hope to open up a debate around Derrida's aporia: if the concept of absolute hospitality is unattainable, at least it keeps us talking about and searching for the most ethical practices in our encounters with strangers. We hope that the chapters in this book, presented alongside a series of photographs by London-based artist Elly Clarke, will open up questions that keep the debates going and that guide us in our search for more ethical answers. The chapters that follow are organized around three core themes, namely locating hospitality in a mobile world, performing the hospitality product, and defining the limits of hospitality. Many of the chapters engage with more than one of these themes as they address the ethics of social relations in a mobile world. Questions surrounding the spatiality and temporality of hospitality, the material and embodied performances of hospitality, and the sometimes hostile negotiations of inclusion and exclusion surrounding hospitable gestures provide thematic linkages throughout the book.

The first three chapters are organized around the spatial and temporal aspects of hospitality encounters. When and where does hospitality happen? How are hospitable places opened up and closed off? Cities, neighbourhoods, cafés, homes and even websites are possible sites of hospitality, where past, present, and future mobilities are welcomed or denied. In Chapter 2, David Bell introduces the concepts of 'host-spots' and 'flickering moments' to draw out the way hospitality is spatially and temporally performed in an urban context. Drawing on a range of examples, from railway catering to London's Olympic bid, and from city cafés to mobile computing, Bell argues that new mobilities and new technologies are challenging assumed dichotomies between public and private and between 'host-at-home' and 'guest-visitor' as the work of hosting or guesting is often performed in mobile and public encounters, such as on trains, on the Web, or in a public square.

In his discussion of hospitality as an element of city promotion, Bell suggests that the unconditional and generous welcome of the hospitable city is often linked to ideas about cosmopolitanism, conviviality, and multiculturalism. Continuing this discussion of urban hospitality, in Chapter 3 Tim Cresswell interrogates the way the city becomes a site of refuge for past migrant mobilities. In this chapter, Cresswell examines the interconnections between place, mobility, memory, materiality and practice in the context of Maxwell Street, Chicago, which has historically been the site of considerable migration both from abroad and internally within the US. Through a close analysis of the documents surrounding local activists' efforts to preserve the area from redevelopment by a nearby university, Cresswell considers the difficulty in institutionalizing sites of memory when those memories are of practices that were necessarily transitory and mobile and left little permanent mark on the material landscape. This chapter thus relocates some of the discussion of hospitality, which is usually used to refer to contemporary migration, by considering the question of whether it is possible to be hospitable to past mobilities.

In Chapter 4, Jennie Germann Molz examines the Internet as a site of hospitality by considering the way virtual and face-to-face hospitality encounters occur on and are organized through social networking websites. Her analysis focuses on hospitality club websites where members offer to host each other as they travel the world. While the websites ostensibly coordinate the logistics of face-to-face hospitality, they also engage the technological interface to coordinate the parameters of reciprocity and reputation through which the hospitality of the web community operates. In this sense, Germann Molz continues the theme of cosmopolitanism introduced in previous chapters, arguing that the hospitality offered between members serves the cosmopolitan fantasy of (safe) proximity to difference, intercultural exchange and an ability to feel at home in the world.

In their individual approaches to the spatial and temporal dimensions of mobility and hospitality, these first three chapters also encourage us to rethink the categories of 'host' and 'guest' by questioning who or what can be a host or guest, and when and where hosting and guesting might happen. As such, they highlight the fluidity of these categories, and challenge the dichotomy between the host and the guest. In particular, they point toward the various ways hospitality is *performed* in space and time.

Place and time are crucial to the production of hospitality, as are the embodied mobility practices and performances of 'hosting work' and 'guesting work' (see Bell, Chapter 2; Veijola and Jokinen, 2005). The next three chapters take up the theme of performing the hospitality product with three empirical studies that draw out the intricate connections between places, bodies, and mobilities in commercial forms of hospitality. These chapters show how basic provisions of hospitality toward the body – nourishment, shelter, and well-being – are implicated in complex performances of sociality, inclusion and exclusion, class, and lifestyle. In Chapter 5, Viv Cuthill explores the intersecting and contested hospitalities and socialities of eating and drinking venues in two northern English towns, Harrogate and Whitehaven. She argues that cafés, bars and restaurants flow in and out of places, signalling messages of style and taste that circulate not only in the way these tourist destinations are imagined, but also in the performance of lifestyle by the social groups who frequent these venues. Moving beyond a dichotomy of production and consumption in defining hospitality, Cuthill proposes that hospitable service cultures are best understood as an embodied and active performance by staff and customers. However, in comparing certain social group mobilities, such as those of tourists and local residents, she also emphasizes that these performances in commercial hospitality spaces represent negotiations of inclusion and exclusion.

In Chapter 6, Tom O'Dell examines how notions of health and well-being circulate in the commercial hospitality site of the spa in Sweden. Whereas many of the chapters in the book focus on the large-scale mobilities of tourism and migration, O'Dell introduces the notion of 'micro-mobilities' to bring our attention to the way hospitality is also orchestrated through small, localized, and embodied movements. In particular, he charts the way bodies move through the spa facility, the way spa treatments are envisioned as mobilizing energy within the body, and the way spa workers' bodies are implicated in the everyday mobility regimes of the spa. His analysis also pays careful attention to links between the micro-mobilities of the spa and broader issues related to class and social trends within Sweden.

In Chapter 7, Paul Lynch, Maria Laura Di Domenico and Majella Sweeney focus on hospitality encounters in commercial homes, where guests pay for accommodation in the host's private home. Here, as the commercial transaction between the host and the guest is negotiated through interpersonal and often friendly relations, the dichotomies between public and private, home and away, guest and host begin to unravel. In place of strict dichotomies, the authors describe a fluid site of hospitality where physical journeys of guests intersect with the metaphorical journeys of hosts to imbue the home with a sense of both permanence and movement. Central to these hospitality performances, they argue, are various material artefacts, including the house itself, that mark home as a dynamic site of welcome familiarity and of worldly travel. In challenging these dichotomies, Lynch, Di Domenico and Sweeney bring our focus to the liminality of hospitality. Performances of hospitality necessarily transgress a threshold, whether this threshold is literally the door of the home or the nation's border, or the figurative threshold between self and other. In such transgressions, the line between hospitality and hostility is constantly in play, underpinning efforts to create openings while simultaneously policing boundaries. The final chapters of the book are concerned with the threshold that epitomizes the hospitality encounter, asking what defines the limits of hospitality.

While many of the previous chapters in the book focus on various forms of tourist mobilities and hospitalities, the last four chapters of the book return to the limits of hospitality in the context of migration and nationalism. These chapters focus on the failure of hospitality to embody its ethical ideal within national contexts. These chapters focus on the ways that gestures of hospitality, welcoming the other, and feelings of belonging, are often highly negotiated and contested through either symbolic or physical violence. With case studies ranging from France and Britain, to the case of queer Russian migrants in Israel and the history of hospitalities in the Middle East, these chapters examine the way differing national discourses of hospitality negotiate the tension between pride, fear, and shame. They highlight how belonging is as inextricably connected to the closures and failures of hospitality as it is to openings. These chapters ably demonstrate how hospitality, welcoming the other and the ability to feel at home, are inseparable from questions of nationality, race, ethnicity, religion, sexuality, class, and gender. It is hospitality encounters, whether public or private, national or individual, that structures these social relations. Developing themes raised throughout the book, these chapters explore the spatial and temporal aspects of hospitality encounters. A key focus examined is the haunting of contemporary hospitality encounters by the past and on the relationship between past and present mobilities and hospitalities.

Similar to Germann Molz's analysis of online hospitality clubs (Chapter 4), Adi Kuntsman charts the overlap between the virtual space of an online discussion forum with the physical space of a South Tel Aviv club in considering the case of queer Russian migrants in Israel. In her discussion in Chapter 8, she demonstrates how these two spaces of hospitality are central to the fierce and sometime violent negotiation of a queer migrant subjectivity and of a sense of belonging. As Jews, these Russian migrants are welcomed by the Israeli 'Law of Return' that defines their immigration as a national home-coming. Yet, this hospitable home-coming is problematized by the simultaneous process of othering and exoticization by Israeli society, and the Israeli queer community. These migrants, as Kunstman argues, who were Jewish in Russia have now become Russian in Israel. Kuntsman traces how religious, national, and sexual identities all intersect with both the opening and closing of hospitality.

This failure of hospitality is also examined in Chapter 9, where Sarah Gibson considers both political and popular discourses of hospitality surrounding the arrival of asylum seekers in Britain. Gibson observes an underlying tension between pride surrounding Britain's identity as a hospitable nation and the nation's increasing hostility directed towards the figure of the asylum seeker. The pride that is invoked in past moments of hospitality is in stark contrast with the shame associated with the securitization of the nation's borders in response to the contemporary 'asylum problem' in Britain. She argues that in contrast to the public rhetoric of 'our' hospitality, the nation's promise of hospitality can never be fulfilled since a nation's hospitality necessarily has to be limited, restricted and conditional. However, this impossibility does not mean that gestures towards hospitality, tolerance and generosity should be abandoned altogether. Instead, Gibson suggests that the underlying ethics of hospitality provide hope for recovering the promise of British hospitality in the future.

This temporal aspect of hospitality is continued in the final two chapters of the volume. These chapters examine how contemporary debates on national identity, citizenship, and belonging in France are inseparable from France's colonial past and discourses of Orientalism. In Chapter 10, Karima Laachir examines the xenophobia and hostility directed towards the figure of the migrant in Europe today. She traces how discourses of hospitality are haunted by the legacy of colonialism with its hierarchical and racist subordination of other cultures and peoples. Hospitality, she argues, is today marked by closure and fear in France. By focusing her attention on the marginalization of postcolonial diasporic communities, Laachir shows the temporal inadequacy of continuing to theorize the 'immigrant as guest' (Rosello, 2001) in contemporary debates of national identity, citizenship, and belonging in France. The 'inclusive exclusion' of these communities within France perpetuates the power structures of colonialism and Orientalism. Laachir argues that Islam has emerged as the cultural and racial other to France's (and Europe's) self. This limit of hospitality is figured through the veiled woman. The veil, as a 'multilayered signifier' (Yegenoglu, 1998: 47), plays a crucial role in articulating sexual and cultural difference in Orientalist discourses.

The discourse of Orientalism is also examined by Judith Still in the last chapter to this volume. In Chapter 11, Still outlines the complexities and contradictory invocations of hospitality during eighteenth century France and examines the differing mobilities and hospitalities embodied in the figure of Jean Chardin, a Protestant diamond merchant who later became one of the Huguenots who sought refuge in Britain from the intolerance of Catholic France. Chardin's life exemplifies how the significance of hospitality varies in its temporal and spatial contexts. As a French diamond merchant, Chardin's writings contributed to the discourse of

Orientalism, and to the representation of Oriental or 'Arab' hospitality. The nomadic ideal of hospitality is imagined in the space of the caravanserai, while the sybaritic hospitality is urban, luxurious, and associated with the harem. These representations of hospitality are in contrast to the French inhospitality (and British hospitality) shown towards the Protestant community during the same period. Still argues that the temporal and spatial aspects of hospitality illuminate the discourse of hospitality in present day France. Hospitality in other times and spaces is central to the analysis and understanding of contemporary discourses and invocations of hospitality. What constitutes hospitality, hospitableness, or welcoming the stranger in a particular moment today can only be understood in relation to these intertexts. This intertextual quality of hospitality is thus repeated throughout all the sections and chapters of this volume. Hospitality, as illustrated through the different practices, spaces, and times encountered throughout the book, is an exceptionally mobile concept in theorizing social relations between people in an increasingly mobile world.

References

Ahmed, S. (2000), *Strange Encounters* (London: Routledge Books).

—— (2004), *The Cultural Politics of Emotion* (Edinburgh: Edinburgh University Press).

Ahmed, S., Castañeda, C., Fortier, A.-M. and Sheller, M., eds (2003), *Uprootings/Regroundings: Questions of Home and Migration* (Oxford: Berg Publishers).

Amin, A. (2004), 'Multi-ethnicity and the Idea of Europe', *Theory, Culture & Society* **21**(2), 1–24.

Appadurai, A. (1996), *Modernity at Large: Cultural Dimensions of Globalization* (Minneapolis: University of Minnesota Press).

Aramberri, J. (2001), 'The Host Should Get Lost: Paradigms in the Tourism Theory', *Annals of Tourism Research*, **28**(3), 738–761.

Aristarkhova, I. (1999), 'Hosting the Other: Cyberfeminist Strategies for Net-Communities' in Sollfrank and Volkart (eds).

—— (2000), 'Otherness in Net-Communities: Practising Difference in Post-Soviet Virtual Context', Paper presented at Situating Technologies Symposium, De Balie, Amsterdam, 1 April 2000.

Bauman, Z. (1997), 'The Making and Unmaking of Strangers' in Werbner and Modood (eds).

—— (1998), *Globalization: The Human Consequences* (Cambridge: Polity Press).

—— (2000), *Liquid Modernity* (Cambridge: Polity Press).

—— (2002), *Living Together in a Full World*, Paper Presented at the On Hospitality Seminar, Leeds University, Leeds, UK, 2–4 March 2002.

Beck, U. (2000), *What is Globalization?* (Cambridge: Polity Press).

Bell, D. (2007), 'The Hospitable City: Social Relations in Commercial Spaces', *Progress in Human Geography*, **31**(1), 7–22.

Benhabib, S., ed. (1999), *Democracy and Difference: Contesting the Boundaries of the Political* (Princeton, N.J.: Princeton University Press Books).

—— ed. (2004), *The Rights of Others: Aliens, Residents and Citizens* (Cambridge: Cambridge University Press).

—— ed. (2005), 'Disaggregation of Citizenship Rights', *Parallax*, **11**(1), 10–18.

—— (2006), *Another Cosmopolitanism: Hospitality, Sovereignty, and Democratic Iterations*. Post, R. (ed.) (Oxford and New York: Oxford University Press).

Bennett, J.B. (2003), *Academic Life: Hospitality, Ethics and Spirituality* (Bolton, MA: Anker).

Brah, A. (1996), *Cartographies of Diaspora* (London: Routledge Books).

Browner, J. (2003), *The Duchess Who Wouldn't Sit Down: An Informal History of Hospitality* (New York: Bloomsbury Publishing).

Caputo, J.D. (1997), *Deconstruction in a Nutshell* (New York: Fordham University Press).

Castells, M. (1996), *The Rise of the Network Society* (Cambridge, MA: Blackwell).

Chan, W. (2005), 'A Gift of a Pagoda, the Presence of a Prominent Citizen, and the Possibilities of Hospitality', *Environment and Planning D Society and Space,* **23**(1), 11–28.

Cheong, S. and Miller, M. (2000), 'Power and Tourism: A Foucauldian Observation', *Annals of Tourism Research*, **27**(2), 371–390.

Choi, J.-G., Woods, R.H. and Murrmann, S.K. (2000), 'International Labor Markets and the Migration of Labor Forces as an Alternative Solution for Labor Shortages in the Hospitality Industry', *International Journal of Contemporary Hospitality Management*, **12**(1), 61–67.

Clarke, N. (2004), 'Mobility, Fixity, Agency: Australia's Working Holiday Programme', *Population, Space and Place*, **10**(5), 411–420.

Clifford, J. (1997), *Routes: Travel and Translation in the Late Twentieth Century* (Cambridge, MA: Harvard University Press).

Cohen, T., ed. (2001), *Jacques Derrida and the Humanities: A Critical Reader* (Cambridge: Cambridge University Press).

Coles, T.E. and Timothy, D.J. (eds) (2004), *Tourism, Diasporas and Space* (london: Routledge Books).

Cresswell, T. (2001), 'The Production of Mobilities', *New Formations*, **43**(1), 11–25.

Crick, M. (1985), 'Tracing the Anthropological Self: Quizzical Reflections on Fieldwork: Tourism and the Ludic', *Social Analysis*, **17**, 71–92.

de Vries, H. (2001), 'Derrida and Ethics: Hospitable Thought' in Cohen (ed.).

Derrida, J. and Düttmann, A.G. (1997), 'Perhaps or Maybe' in Dronsfield and Midgley (eds).

Derrida, J.(ed.) (1999), *Adieu to Emanuel Levinas*. trans. Brault, P.-A. and Naas, M. (Stanford, CA: Stanford University Press Books).

—— (2000), *Of Hospitality, Anne Dufourmantelle Invites Jacques Derrida to Respond*. trans. Bowlby, R. (Stanford: Stanford University Press Books).

_____. (2001a), *Deconstruction Engaged,* P. Patton and T. Smith (eds) (Sydney: Power Publications).

—— (2001b), *On Cosmopolitanism and Forgiveness*. trans. Rorty, R. (London and New York: Routledge Books).

—— (2003), Interviewed in G. Borradori, *Philosophy in a Time of Terror* (Chicago, IL: University of Chicago Press).
—— (2005), 'Principle of Hospitality', *Parallax*, **11**(1), 6–9.
Dikeç, M. (2002), 'Pera Peras Poros: Longings for Spaces of Hospitality', *Theory, Culture & Society*, **19**(1–2), 227–247.
Dronsfield, J. and Midgley, N. (eds) (1997), 'Responsibilities of Deconstruction', *PLI – Warwick Journal of Philosophy*, 1–18.
Duval, D.T. (2003), 'When Hosts Become Guests: Return Visits and Diasporic Identities in a Commonwealth Eastern Caribbean Community', *Current Issues in Tourism*, **6**(4), 267–308.
—— (2004), 'Mobile Migrants: Travel to Second Homes', in Hall and Meuller (eds).
Feng, K. and Page, S.J. (2000), 'An Exploratory Study of the Tourism, Migration-Immigration Nexus: Travel Experiences of Chinese Residents in New Zealand', *Current Issues in Tourism*, **3**(3), 246–281.
Foucault, M. (1986), 'Of Other Spaces' trans. J. Miskowiec, *Diacritics*, **16**(1), 22–27.
Friese, H. (2004), 'Spaces of Hospitality' trans. J. Keye, *Angelaki*, **9**(2), 67–79.
Gibson, S. (2003), 'Accommodating Strangers: British Hospitality and the Asylum Hotel Debate', *Journal for Cultural Research*, **7**(4), 367–386.
—— (2006), '"The Hotel Business is About Strangers": Border Politics and Hospitable Spaces in Stephen Frears's *Dirty Pretty Things*', *Third Text*, **20**(6), 693–701.
Gilroy, P. (1993), *The Black Atlantic* (London: Verso Books).
Grace, H., Hage, G., Johnson, J., Langsworth, J. and Symonds, M. (1997), *Home/World: Space, Community and Marginality in Sydney's West* (Annandale, N.S.W.: Pluto Publishing).
Hage, G. (1997), 'At Home in the Entrails of the West' in Grace et al.
—— (2002), '"On the Side of Life": Joy and the Capacity of Being' A Conversation with Ghassan Hage' in *Hope: New Philosophies for Change*. Zournazi, M. (ed.) (London: Lawrence & Wishart Books).
—— (2003), *Against Paranoid Nationalism: Searching for Hope in a Shrinking Society* (London: Merlin Press Books).
—— (2005), 'Nomadic Hospitality and the Gift of Rest', Paper presented at Mobilizing Hospitality: The Ethics of Social Relations in a Mobile World Workshop , Lancaster University, UK, 26–27 September 2005.
Hall, C.M. and Jenkins, J. (1995), *Tourism and Public Policy* (New York: Routledge Books).
Hall, C.M. and Mueller, eds (2004), *Tourism, Mobility and Second Homes* (Clevedon: Channel View Publications).
Hall, C.M. and Williams, A.M., eds (2002), *Tourism and Migration: New Relationships between Production and Consumption* (Dordrecht: Kluwer Academic Publishers).
Hannam, K., Sheller, M. and Urry, J. (2006), 'Editorial: Mobilities, Immobilities and Moorings', *Mobilities*, **1**(1), 1–22.

Harding, J. (2000), *The Uninvited: Refugees at the Rich Man's Gate* (London: Profile).

Heal, F. (1990), *Hospitality in Early Modern England* (Oxford: Oxford University Press).

Hetherington, K. (1997), 'In Place of Geometry: the Materiality of Place' in Hetherington and Munro (eds).

Hetherington, K. and Munro, R., eds (1997), *Ideas of Difference* (Oxford: Blackwell Books).

Heuman, D. (2005), 'Hospitality and Reciprocity: Working Tourists in Domenica', *Annals of Tourism Research* **32**(2), 407–418.

Honig, B. (1999), 'Difference, Dilemmas, and the Politics of Home' in Benhabib (eds).

Joly, D. and Cohen, R. (1989), *Reluctant Hosts: Europe and Its Refugees* (Aldershot: Avebury).

Kant, I. (1996 [1795]), 'Toward Perpetual Peace: A Philosophical Project' in *Practical Philosophy: The Cambridge Edition of the Works of Immanuel Kant*. Gregor, M.J. (trans.) (Cambridge: Cambridge University Press).

Kaufman, E. (2001), *The Delirium of Praise* (London: Johns Hopkins University Press).

Kelly, E. (2006), 'White Hospitality: A Critique of Political Responsibility in the Context of Australia's Anti-Asylum-Seeker Laws', *Continuum: Journal of Media and Cultural Studies*, **20**(4), 457–469.

Lashley, C. and Morrison, A., eds (2000), *In Search of Hospitality: Theoretical Perspectives and Debates* (Oxford: Butterworth Heinemann).

Lashley, C., Lynch, P. and Morrison, A., eds (2007), *Hospitality: A Social Lens* (Oxford: Elsevier Books).

Lynch, P. (2005), 'Sociological Impressionism in a Hospitality Context', *Annals of Tourism Research*, **32**(3), 527–548.

Massey, D. (1994), *Space, Place and Gender* (Cambridge: Polity Press).

McNaughton, D. (2006), 'The "Host" as Uninvited "Guest": Hospitality, Violence and Tourism', *Annals of Tourism Research*, **33**(3), 645–665.

Metselaar, S. (2005), 'When Neighbours Become Numbers: Levinas and the Inhospitality of Dutch Asylum Policy', *Parallax*, **11**(1), 61–69.

Morley, D. (2000), *Home Territories: Media, Mobility and Identity* (London: Routledge Books).

Nussbaum, M.C. (1994), 'Patriotism and Cosmopolitanism', *Boston Review*, **19**(5), 3–34.

O'Reilly, K. (2003), 'When is a Tourist?, The Articulation of Tourism and Migration in Spain's Costa del Sol', *Tourist Studies*, **3**(3), 301–317.

Perera, S. (2002), 'A Line in the Sea', *Race and Class*, **44**(2), 25–39.

Pratt, M.L. (1992), *Imperial Eyes: Travel Writing and Transculturation* (London: Routledge Books).

Puar, J.K. (2002), 'Circuits of Queer Mobility: Tourism, Travel, and Globalization', *GLQ*, **8**(1–2), 101–137.

Pugh, M. (2004) 'Drowning Not Waving: Boat People and Humanitarianism at Sea', *Journal of Refugee Studies*, **17**, 50–69.

Pugliese, J. (2002), 'Penal Asylum: Refugees, Ethics, Hospitality', *Borderlands* **1**(1), available online at: <http://www.borderlandsejournal.adelaide.edu.au/vol1no1_2002/ pugliese.html> (accessed: 10.11.06).

Rosello, M. (2001), *Postcolonial Hospitality: The Immigrant as Guest* (Stanford, CA: Stanford University Press Books).

Savic, O. (2005), 'Figures of the Stranger: Citizen as a Foreigner', *Parallax*, **11**(1), 70–78.

——, ed. (2005), 'Seeking Asylum', Special issue of *Parallax*, **11**(1).

Schlunke, K. (2002), 'Sovereign Hospitalities?', *Borderlands* **1**(2), available online at: http://www.borderlandsejournal.adelaide.edu.au/vol1no2_2002/schlunke_hospitalities.html (accessed: 10.11.06).

Selwyn, (1996), *The Tourist Image: Myths and Myth Making in Tourism* (New York: John Wiley & Sons Books).

—— (2000), 'An Anthropology of Hospitality' in Lashley and Morrison (eds).

Sherlock, K. (2001), 'Revisiting the Concept of Hosts and Guests', *Tourist Studies*, **1**(3), 271–295.

Simmel, G. (1971), *On Individuality and Social Forms.* Levine, D.N. (ed.) (London and Chicago: University of Chicago Press Books).

Skeggs, B. (2004), *Class, Self, Culture* (London: Routledge Books).

Smith, V., ed. (1989), *Hosts and Guests: The Anthropology of Tourism*, 2nd edn (Philadelphia, PA: University of Pennsylvania Press).

Sollfrank, C. and Volkart, Y., eds (1999), *Next Cyberfeminist International Reader* (Hamburg: Old Boys Network) [website], <http://www.obn.org/obn_pro/downloads/reader2> (accessed: 12.08.05).

Still, J. (2004), 'Language as Hospitality: Revisiting Intertextuality via Monolingualism of the Other', *Paragraph*, **27**(1), 113–127.

—— (2006), 'France and the Paradigm of Hospitality', *Third Text*, **20**(6), 703–710.

Tomlinson, J. (1999), *Globalization and Culture* (Cambridge: Polity Press).

Tribe, J. (2005), *The Economics of Leisure and Tourism*, 3rd edn (Oxford: Butterworth-Heinemann Books).

Urry, J. (2000), *Sociology Beyond Societies* (London: Routledge Books).

Veijola, S. and Jokinen, E. (2005), *'Hostessing, Gender and Work'*, paper presented at special session 'Hostesses of the World: Sexuality, Power and Gender', 37th International Conference of the International Institute for Sociology (IIS), Stockholm, Sweden, 5–9 July 2005.

Venn, C. (2002), 'Altered States: Post-Enlightenment Cosmopolitanism and Transmodern Socialities', *Theory, Culture & Society*, **19**(1–2), 65–80.

Vertovec, S. and Cohen, R., eds (2002), *Conceiving Cosmopolitanism* (Oxford and New York: Oxford University Press).

Waldron, J. (2006), 'Cosmopolitan Norms' in Benhabib, S. (ed.) (Oxford and New York: Oxford University Press).

Werbner, P. and Modood, T., eds (1997), *Debating Cultural Hybridity* (London: Zed Books).

Worth, H. (2006), 'Unconditional Hospitality: HIV, Ethics and the Refugee "Problem"', *BioEthics*, **20**(5), 223–232.

Wyschogrod, E. (2003), 'Autochthony and Welcome: Discourses of Exile in Levinas and Derrida', *Journal of Philosophy and Scripture*, **1**(1), 36–42.

Yegenoglu, M. (1998), *Colonial Fantasies* (Cambridge: Cambridge University Press).

—— (2003), 'Liberal Multiculturalism and the Ethics of Hospitality in the Age of Globalization', *Postmodern Culture*, **13**,(2), available at <http://www3.iath.virginia.edu/pmc/text-only/issue.103/13.2yegenoglu.txt> (accessed: 15.03.06).

Seville to Hackney: A Photographic Journey

Elly Clarke

In Seville, a guide and tourists looking up. The beginning of the film.
In Vienna, an afternoon sleep, my lover is in the bathroom.
In Tijuana, warning posters.
In Zacatecas, songs from a man with few teeth, but a voice that would carry across valleys.
Back at the hotel, washing up to dry.
In Chiang Mai, slow connections.
In Guadalajara, a shabby bar, just closed, or just about to open, promises a good time.
On the way to China from Laos, ladies offer drugs in little waxed paper sachets, hidden beneath hand-made bracelets.
At home, on my sofa in Hackney. The table I bought for £15 from Brick Lane.
At The Last Resort, on Ko Lanta, stuck to the fake wood ceiling, attracted to the light, a ghecko lets out its bird-like call.
Crossing borders.
Some are easier than others.
For some, some may be easier than others.
Into Laos, polysterene packaging is pushed in.
In Luang Prabang, threadbare glamour is mine for $3 a night. The sounds of the night enter through the slatted walls.
Back in Hackney, I pack up my room.

Elly Clarke is based in London. She is interested in the impact of mobility on people's sense of self, both when alone and part of a community. As a photographer, Clarke's projects are often participatory, where, through the distribution of cameras to other people, she is able to illustrate multiple perspectives of a given situation, event or community, from the inside as well as out. Her most recent participatory project was one she did on the Trans-Siberian Train, from Moscow to Beijing, where she conducted interviews with and gave cameras to Russian, Chinese and Mongolian passengers on the train to learn more about the journey and the cultures of the countries she was passing through, than she could have done by herself, as a lone traveller. This project was also about overcoming language barriers.

More info: http://www.axisweb.org/artist/ellyclarke

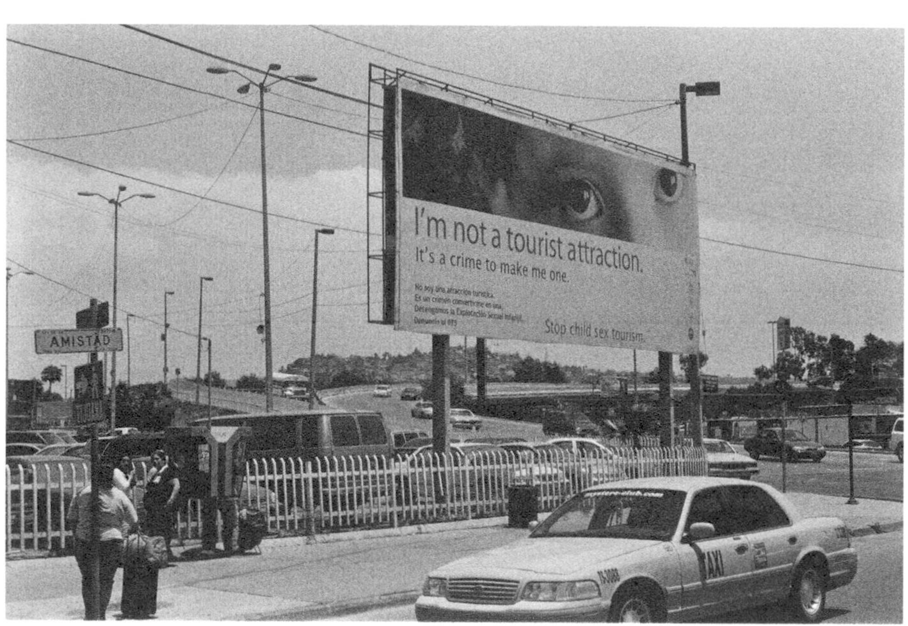

The Limits of Hospitality **(Tijuana, Mexico, 2006) by Elly Clarke**

Chapter 2

Moments of Hospitality

David Bell

As it is most commonly conceived and thought with, the relationship between host and guest that defines hospitality is marked by an asymmetry of mobility. The host is static, fixed, rooted, while the guest is footloose, on the move, rootless. This asymmetry defines the very 'host-ness' of the host and the 'guest-ness' of the guest. The host is at home, either literally in his house or more broadly in his homeland; the guest is an incomer, a visitor, a stranger.[1] While Derrida (2001) and others have spilt much ink playing with the instabilities and double entendres latent in this host-guest formula, it is safe to say that the relationship is defined by mobility. Indeed, the guest's mobility is implied in the welcome he is given: he is hosted on the understanding that he will not outstay his welcome, that he will move on, and on. And, of course, temporality marks the stability and mutual exclusiveness of these identities, for when a guest settles and becomes rooted, he can become a host; and when a host leaves home to travel, he transforms into a guest (though each carries with them traces of their former identity).

This commutation, I want to argue in this chapter, has become even more rapid and pervasive, for reasons that I shall elaborate, further destabilizing the host-guest relation, leaving the identities of host and guest fragile, uncertain, decentred. As Mimi Sheller and John Urry (2004: 8) write, in the context of tourism: 'many "hosts" are increasingly also from elsewhere', are also on the move, passing through, guests enacting host-ness. Hospitality – as a relationship marked by poles of host-ness and guest-ness, and by the obligations and rewards that this bipolarity brings – is thus itself destabilized as we enter an increasingly mobile age, a society of mobilities (Sheller and Urry 2006; Urry 2000).

Of course, this new vector of mobility is unevenly developed and inequitably shared, such that some groups experience only further immobility – especially, perhaps, in terms of the ability of cross borders, to make a new home in another place. What we might call 'mobility capital' and the 'mobile habitus', to twice borrow from Bourdieu (1984), have become new indexes of status, new markers of distinction. But not all forms of mobility are tradable; some people remain marked by their 'mobile-ness', their guest-ness, or their not-host-ness. We should remember the relation between physical and social mobility here, too: the prestige of hosting is an unequal resource (though hosting can be a heavy burden as well as a privilege,

1 The gendered pronouns are used here purposely – the formula I am thinking about here implies a male host, 'man of his house', who can offer hospitality which, we can presume, is the product of (gendered) domestic labour, rather than *his* own labour (Bell, 2007).

just as the status of guesting is context-specific). And in a society marked by what Andreas Wittel (2001) calls a 'network sociality', we need to rethink how hospitality happens in what Sheller and Urry (2006: 222) describe as 'flickering moments' or 'networks of hosts-guests-time-space-cultures' that come together contingently (Sheller and Urry, 2004: 6).

It is these moments that I want to think about in this chapter, through searching for and trying to understand what we might refer to – tweaking a familiar term from wi-fi technology, that great enabler of network sociality – as 'host-spots': more or less stable or fragile places and/or times when hosting-guesting occurs, or when host-like and guest-like potentialities are afforded. As Sheller and Urry (2006: 214) put it, these host-spots, these moments of hospitality, are 'not so much fixed as implicated within complex networks by which hosts, guests, buildings, objects and machines are contingently brought together to produce certain performances in certain places and certain times'. To narrow the focus down, I will concentrate my discussion on hospitality in an urban context – as part of my on-going work on 'the hospitable city' (see also Bell, 2007). As I will go on to show, the contingent performances and comings-together of hosts and guests in cityspaces unsettle our understandings of hospitality, offering an opportunity to think it differently. I will do so through a discussion of selected 'moments' of hospitality: mobile host-work in the context of commuting travel, mega-events and hospitality, everyday urban hospitableness, and 'techno-mobile' hosting and guesting.

Mobile Hosts

In the language of networked, wireless computing, a 'mobile host' is a computational device equipped to connect to networks (notably the Internet) while on the move, yet still using its 'home address' – aided by a protocol that facilitates 'seamless' movement through and between networks, such that the user is oblivious to the complex connection-work taking place on their behalf (Perkins, Myles and Johnson, 1994). The term comes to us, therefore, with some preset connotations, which I would like to hold on to and revisit later, while also thinking about the impact of the 'new mobilities paradigm' on the role of host and the possibilities of hosting, of performing host-ness. I begin with a mundane and familiar instance of mobile hosting: I want to think about those people whose job entails providing hospitality services on passenger transport systems. For example, in the UK, following denationalization, railway companies have undergone a thoroughgoing 'customerization' process, a process which has redefined work roles for many railway workers. As is commonplace in such work-role redefinitions, jobs have been renamed to better reflect this new customer service ethos, so the train guard became the train manager and so on. Catering services on some rail networks were put out to competitive tender in the wake of denationalization, with companies such as Rail Gourmet (a sister company of the airline caterer, Gate Gourmet, and part of a larger European company with a diverse portfolio) winning contracts to provision travellers (sorry, 'customers').

During a period of regular rail commuting, I became fascinated by this form of mobile hosting, observing Rail Gourmet 'hosts' or 'stewards' in their daily working lives, getting on and off particular trains, with their specially-designed trolleys piled with drinks and snacks, negotiating the metal ramps they use to bridge the gap at the platform edge. Of course, these train-bound providers are close relations to the more thoroughly researched airline stewards, the 'trolley dollies' made famous for their 'emotional management' and tightly-controlled work performances (Hochschild, 1983). While seemingly less governed by such 'emotion rules', these mobile hosts are nonetheless tasked with a particular performance of commercial hospitality – providing food and drink – as they experience the motion of train travel and the motion of constantly changing guests, as travellers board and de-board. As subcontracted hosts, however, they are not direct employees of the rail company, yet they often share the labours of on-train hosting, giving travel information, helping passengers with bags and generally participating in the social life of commuting.

Rail commuting, in fact, produces a complex network of performances of hosting and guesting, between railway staff (and subcontracted staff) and travellers, and also between travellers (asking to share such proximal seating space echoes a host-guest relation, as some travellers signal 'ownership' of travel-space, and thus host-ness, through bodily comportment, deployment of belongings, and so on – a point I shall return to later). So, while the hubs, nodes or 'still points' of mobile society, such as airports or train stations, have been thought through in terms of their role in 'network hospitality', the transport containers themselves, and the various people and objects they carry, are also implicated in the production of mobile host-ness and guest-ness. The post-9/11 presence of air marshals on US planes, and in the UK the heightened visibility of British Transport Police post-7/7, adds a further dimension to mobile hosting and surveilling guests (as well as promoting self-surveillance and docility, not to mention the self-conscious performance of non-threatening 'not-terrorist-ness').

Of course, the host-guest moments that flicker by as the train shuttles its passengers from A to B and back again is but one part of a bigger set of mobility stories, including the story of the durability of commuting as a work-life experience, despite the predictions of a near future of telecottaging and virtual workplaces. The network society is still dominated by cities, even as urban forms mutate and morph, as cities become 'cybercities' (Graham, 2002), as the 'space of flows' and the 'space of places' exhibit copresence, albeit with a complex layering (Castells, 1996). The survival of the city as a form of social-spatial organization is familiarly periodized, with the present labeled postmodern or post-industrial, marking a shift in the functions of cities under changing global economic conditions. Post-industrial cities become spaces of consumption and spectacle, attractors of footloose global flows, command centres of the new economy, key symbolic sites, and destinations for all sorts of flows of different people and things (Short, 2000). Central to the post-industrializing of cities has been a new urban competitiveness, built on the promise of a loosening of the 'old' urban hierarchy, with 'new' opportunities for cities to sell themselves and to draw down the valued resources of the post-industrial era. This intense inter-urban competition can be vividly witnessed in the contest to become host city to global events – hence our second moment of mobile hosting, the hospitality event.

The Hospitality Offer

As Maurice Roche (2000) extensively documents, 'mega-events' have a long and important history in the fortunes of cities (and nations). World's fairs and expos, global sporting events, and the legion of conventions, festivals and assorted comings-together have become increasingly prized in an era of inter-urban competition, not only for the supposed (direct and indirect) economic benefits they bring, but also for the less measurable, 'softer' social, cultural and symbolic benefits. A whole industry of promotion, selection and accounting has grown up around such mega-events, and some cities have developed the self-promotional savvy to bid successfully for, and then capitalize fully on, such temporary hosting (Cochrane, Peck and Tickell, 1996). Hospitableness is here figured not only in terms of an appropriate infrastructure, with bespoke facilities and guaranteed standards, but also in terms of the welcome that 'the city' promises its guests. In the script of successful urban hospitality, people are a key resource, therefore, and the countless small moments of hospitality that must be (officially and unofficially) enacted are an important guarantor of the guesting experience – an experience which brings a supposed cash dividend to the host city, topped off with longer term 'legacy benefits' that will badge the city as hospitable long after the event. The promotional materials for London's upcoming hosting of that most mega of mega-events, the Olympic Games, for example, trades on (selected) cultural characteristics of the city as evidence of the welcome offered to participants and spectators at the Games:

> London has always been one of the world's great cosmopolitan cities. Throughout history, people have come from every continent and corner of the globe. To live, to visit, to mix. The blend doesn't just reflect modern London. It is modern London. Today the city brings together more than 50 ethnic communities of 10,000 or more people. More than 70 different national cuisines are available. And a staggering 300 different languages are spoken. This diversity characterizes the city. Wherever you go you find the vibrancy and variety of integrated communities living together. There's the Afro-Caribbean centre of Brixton, a bustling Chinatown in central Soho, 'Banglatown' in the East End's Brick Lane and a 'little India' in Southall to the west, to name but a few. Throughout the city, citizens from every corner of the world flourish as neighbours. … London loves to celebrate its multiculturalism. Ethnic festivals, events and ceremonies help everyone share and enjoy… In 2012, our multicultural diversity will mean every competing nation in the Games will find local supporters as enthusiastic as back home. In every one of those 300 languages there undoubtedly exists words of encouragement. In London you'll be able to hear them all. As part of London's 2012 offer, there was a promise that the city will come alive as never before 'with energy, passion and excitement'.
>
> (http://www.london2012.org)

The actual metrics of successful hosting are undeniably more complex than such boosterist rhetoric suggests, with many intangibles having to be extrapolated or guesstimated (Shibli and Gratton, 2001); nevertheless, a successful marshalling of the city's hospitality offer can be made to do important symbolic work in reforming a city's image, as iconic examples such as Glasgow post-City of Culture or Manchester's constant bidding to host mega-events attest, albeit not without being contested (Cochrane, Peck and Tickell, 1996; Richards, 2000).[2]

City imagineering places a high premium on attracting host-city status for mega-events, and touts the 'reputation effects' that accrue from successful hosting (the disbenefits of unsuccessful hosting or failed bids are silently brushed aside). Essentialized attributes of the host population are converted into tradable capital in a contest to prove 'host-ness', moreover, while the infrastructures of hospitableness are variously strengthened, polished, repurposed and incentivized. Such processes do, of course, also bring to the surface contradiction and conflict, the 'other cities' that uneasily co-exist alongside the image of the hospitable urban host. Harvey Newman (2000), for example, outlines the tensions between the promotion in the mid-1970s of Atlanta, Georgia, as a 'convention city' and its simultaneous notoriety as the 'murder capital' of the USA. While the very success of Atlanta as a tourist destination heightened anxieties about crime, especially in the downtown area where convention-goers milled, any attempt to cover over negative discourses with positive ones only served to propagate counter-discourses, unsettling any attempt to smoothly brand the city. Similar ambivalences occur in host cities which have to manage public order at the same time as offering a warm welcome, perhaps most notably seen in soccer mega-event hosting, where anxieties over 'hooliganism' clash with the discourse of friendly hospitableness, or in the hosting of political gatherings that bring with them other, unsanctioned counter-gatherings, as in the protests in cities hosting the World Trade Organization.

Aside from mega-event hosting, cities have to perform host-ness in a variety of other ways, too. The fact that they are compelled into this performance is important, by the way: no city would want to brand itself as *inhospitable* – though it may choose to make itself over as very selectively welcoming, and therefore inhospitable to 'others' whose presence is not valued. Discourses of multiculturalism and cosmopolitanism, for example, trade on a welcome to 'diversity', but simultaneously an unwelcome to those whose difference is seen as antithetical or problematic, such as the 'racist' or the 'chav' (Binnie et al., 2005); Skeggs, 2005). The calculus of urban hospitableness has, as noted, become a complex enterprise, as multiple performance indicators, outcomes and outputs are taken as signs of host-ness. New ways of assessing and performing host-ness constantly arise as inter-urban competitiveness remains such a charged game. Arguably the most famous, widely talked-up and most in-vogue manifestation of this currently doing the rounds comes from the American writer Richard Florida (2002, 2005), and his indexing of the 'creative class' and of 'creative

2 In the context of London's Olympic hosting, there are two important points to think about: first, the extent to which the area of the city designated as host-site for the Games has a long history as a host-site for migrant communities, who may feel much less welcome as the Olympics demand a different order of hosting and guesting; second, London's successful bid to host the Olympics came the day before the 7/7 bombings, an event which dramatically rewrote the city's hospitableness, leading to heightened security and surveillance.

cities' – cities performing a particular version of host-ness, aimed at attracting and retaining a particular class of guests.

The Hospitable Class?

It is no understatement to say that Florida's 'creative class' thesis has had (and continues to have) a huge impact on urban governance and imagineering. His focus on how certain cities have attracted high-status 'new' economic sectors – the much-hyped creative industries – has been seized upon by urban managers and boosters, and countless cities have engaged Florida's formula, and his consultancy services, to undertake 'hipsterization strategies' (Peck, 2005: 740) in an effort to become host-spots for the new saviours of cities – creatives, gays and geeks. In Florida's success story for cities, gays and geeks (IT workers) act as an 'advance guard', and have a vital 'hipsterization' function to perform that, if successful, can be traded to attract the real treasure: the creative class. These fickle, funky, footloose 'no collar' workers (Ross, 2003) have particular lifestyle requirements that shape their relocation decisions, and Florida suggests that cities should, indeed must, propagate the requisite amenities favoured by these picky, high-value would-be in-migrants. Central to the Florida formulation is the vision of a trendy, buzzing, happening neighbourhood, an 'edgy' city with a 'people-climate' that frees the mind and gets the creative juices flowing (Peck, 2005). As especially welcomed and valued guest-workers, then, the creative class is seen by Florida as having the power to make or break cities, and the muscle to make its own tastes and preferences dominant. Cities can hardly opt out of the chase, therefore, hence the ubiquity of 'creative cities' initiatives; again, what city would want to sell itself as unwelcoming to this new economic panacea, or as *uncreative*?

Undergirding this creative city, therefore, is a vision of urban hospitality customized to suit the (imagined) needs of the creative class. While the actual employment conditions and working lives of 'creatives' are often marked more by casualization and job insecurity than glamour and 'buzz', thereby casting doubt on the image of the assertive, choosy creative class that Florida provides (see for example: McRobbie, 2002), much less has been said about what Jamie Peck (2005: 756) calls the 'contingent economy of underlaborers' who service the creative class. This economy includes those who provide 'upstream' and 'downstream' goods and services to creatives (Pratt, 2002), and also those who staff the trendy bars, boho boutiques and buzzy eateries that creatives are purported to frequent for socializing and networking.

Research on Manchester, UK by Mark Banks (2006) and on Auckland, New Zealand by Alan Latham (2003), pays some attention to the broader 'hospitality investment' that surrounds the creative class, however. Banks shows how creative workers in Manchester's Northern Quarter were concerned to embed themselves in the locality where they lived and worked, and attempted to foster a broader sense of engagement with the Northern Quarter's heterogeneous communities – the communities that in part attracted creatives to that neighbourhood in the first place. Key 'great good places' (Oldenberg 1999) in the Northern Quarter, such as delis, pubs and cafés, were important sites for this 'community work', as Latham also found in the gentrifying Auckland neighbourhood where his research was based. There, café and bar owners played a central role in defining and promoting the 'feel' of the place, suggesting a more proactive role by local hospitality entrepreneurs in configuring certain performances of 'host-ness' (and 'guest-ness'). As Latham writes, many of the key players in his study area saw what they were doing as 'a kind of socio-cultural project' (2003, 1717) – they were investing in producing new ways of living (and not just new markets for their food and drink). The entrepreneurs creating hospitality spaces in this area of Auckland have played a crucial role in defining the 'feel' of the neighbourhood, Latham argues, consciously shaping their bars and cafés to promote particular kinds of conviviality, informality and hospitableness.

While employment conditions in the hospitality industry – including the requirement or obligation placed on workers to perform in particular stylized ways in order to perpetuate the ambience of a bar or café – have been rightfully critiqued (for example Crang, 1994), the hosting and guesting work that takes place in commercial hospitality spaces should nevertheless be credited for its role in the production and promotion of convivial or hospitable ecologies of place. Here again, of course, there are tensions – one person's convivial ecology is another's noise pollution, and neighbourhoods like Manchester's Northern Quarter or London's Soho have had to manage the sometimes conflicting lifestyles of their residents and visitors (Bell and Binnie, 2005; Roberts and Turner, 2005). As we shall see, moreover, consumption spaces like cafés have become somewhat overburdened with an expectation that the hospitality they provide is a magic solution to revitalizing cities. This issue notwithstanding, work such as that by Latham clearly shows the potential for commercial hospitality spaces to contribute to the 'feel' of their locality, and hence their broader 'hospitality effects'.

Looking even more closely (and ethnomethodologically) at what goes on in cafés, Eric Laurier and Chris Philo (2006) discuss what they call 'the welcome of customers' – the micropractices by which customers as well as workers perform 'hosting' in commercial hospitality spaces, for example through their reactions to each other, their willingness (or not) to share space or assist one another; like the rail commuters mentioned earlier, guests perform host-ness through their interactions, and in so doing they too work to define the 'feel' of a place. Performances of 'regular-ness' or 'local-ness' – a confident comportment, chit chat with staff and with other regulars or locals, and so on – serve to include and to exclude, while commercial exchange practices such as buying rounds of drinks or running a bar tab confer membership selectively (see Laurier, Whyte and Buckner, 2001). In these and countless other ways, moments of hospitality are enacted, successfully or unsuccessfully, inclusively or exclusively, in commercial hospitality spaces. They are also enacted, of course, in other spaces of encounter in cities, including public spaces.

Public Hosting and Guesting

Urban policy-makers in the UK (and elsewhere) have been voicing concern over the state of our public spaces for some time now. Following Richard Rogers' (1999) highly influential *Urban Renaissance* report, they have tended to fetishize public spaces and concomitant public cultures found (or imagined) in certain 'European' or 'Mediterranean' cities, while conversely seeing British urban public spaces as either overly-privatized, too risk-filled, or evacuated by sundry processes of counterurbanization and ghettoization. While there is much to contest in both the valuing of 'European' public culture supposedly afforded by piazzas, street cafés and traditions of using public space, and in the devaluing of the UK's public spaces as either empty, litter-strewn and threatening or as overly surveilled and privatized, this turn to rethinking public spaces (as well as re-planning and redesigning them; see Montgomery, 1995, 1997), has led to more focused attention being paid to the actual practices that do, or might, take place in such cityspaces. Ash Amin (2006: 1019), for example, discusses the potential 're-enchantment' of the 'good city', finding exciting possibilities in 'a certain kind of sociality that comes from particular forms of gathering in public spaces' – gatherings such as 'associations, clubs, car-boot sales, restaurants, open spaces, bolt-holes, libraries, formal and informal gathering-places, and [the] multitude of friendship circles that so fill cities'. Crucially, Amin's vision of this 'good city' sees civic spaces and new public cultures coalescing out of 'free engagement and visibility among strangers in the city's public spaces' (1019). He is not alone in drawing up such a wish-list, and he is also among prominent company in his prognosis that 'urban association is increasingly defined by spectacle and consumption' (1020) and that this will not necessarily produce the forms of public culture he sees as central to 'good' urban re-enchantment (though this is a point I would want to contest: commodified and 'spectacularized' spaces can and do produce positive public cultures and should not be so easily dismissed).

Like Amin, Quentin Stevens (2004) is interested in urban public culture; he chooses to form his wish-list around playful encounters between strangers in cityspaces, mapping the possibilities for play as 'taking one beyond habitual patterns of social engagement' (139). In unpacking urban playfulness in this way, through a discussion of assorted moments of often spontaneous interaction, Stevens provides a fine-grained and engaging discussion of how bodies-in-play – including those who witness playful acts – potentially rewrite or reorder urban public spaces, offering novel ways to engage with those spaces and with the strangers encountered there. While acknowledging limitations and conflicts, and thereby not simply celebrating free play, Stevens provides a very helpful and insightful series of observations about 're-enchantment'. Yet, like Amin, he prefers to focus on happenstance, chance meetings, unplanned and improvised events, implicitly challenging what is assumed to be a prevalent and hegemonic urban ordering and regulatory regime – note his comment on moving 'beyond the habitual'.

In side-stepping more routinized, regulated (and commercialized) spaces, writers like Amin and Stevens neglect much of the everyday 'work' of interaction, of hosting and guesting, that produces and reproduces public cultures of hospitality precisely in habitual ways. While this interaction-work has been explored in the context of tourism (Sheller and Urry, 2004), in other branches of the commercial hospitality industry it is often disregarded as nothing more than cynical, 'calculative' hosting – workers knowingly perform 'host work' because their jobs and wages depend on it (Bell, 2006). This leaves open the question of 'guest performance', of course: tourists and customers also perform in certain ways, perform 'guest work', though their motives are irreducible to purely calculative, commercial logics (Edensor, 2000). These guest performances are equally important in defining a hospitable encounter, though they have been neglected outside of tourism studies, as work on 'interactive service encounters' has tended to focus on the scripting of 'host' performances, overlooking the ways customers or clients perform, too (Du Gay, 1996).

However, what I am more interested in here, to return to Sheller and Urry's (2006) evocative phrase, is those 'flickering moments' of hosting and guesting that occur between strangers in urban public spaces – and, of course, in commercial and private spaces, too. In an era of heightened mobility (for some, at least), these flickering moments in fact rewrite the public and the private, folding or collapsing them together; publics and privates, as Sheller and Urry (2003: 108) put it, are increasingly 'being performed in rapid flashes within less anchored spaces' – mobile phone users, for example, hold 'private' conversations in public, while the car is a private 'capsule' moving along public roads (yet, at the same time, these technologies complexly reconfigure, de-differentiate and thoroughly mix-up the very meanings of public and private; see Sheller, 2004). So, in this context, what does it mean to speak of strangers in public spaces performing hosting and guesting with each other?

The flickering moments, the rapid flashes, not only reconfigure public and private; they reconfigure host and guest, too. In his discussion of definitions of hospitality, Conrad Lashley (2000) differentiates social, private and commercial hospitality domains, mapping out their different modes of hospitality (and also their convergences and overlaps). Private hospitality takes place in the home, commercial hospitality in economic exchange contexts, while social hospitality for Lashley means the broader social codes, forms of sociality and sociability, that can be enacted in private or commercial spaces. Social hospitality also hints at social inhospitality, as in consumption-based status differentiations that are deployed to mark social class location, for example (Bourdieu, 1984; Skeggs, 2005). Yet social hospitality (and inhospitality), I would argue, also occurs between people in public space; not only between those marked and readable as 'hosts' and 'guests', for example in a tourist site, but also in performances of mutual hosting and guesting witnessed in mundane, largely gestural, public encounters. Behaviour in public, as Goffman (1963) classically elaborated, is composed of myriad gestures, particular modes of comportment, ways of occupying and moving through space, glances, nods and winks. Just as customers and workers in the café Laurier and Philo (2006) observed collectively performed hosting and guesting, so the many 'routine encounters and shared experiences' in urban public spaces underpin what Paul Knox (2005: 8) describes as 'the intersubjectivity that is the basis for both a sense of place and for a structure of feeling within a community'. Pursuing this intersubjective production of hospitality in his study of gay villages and Chinatowns, Vince Miller (2006) uses the term 'intimacy' in order to try to capture the 'felt aspect of belonging or not belonging' which he rightly argues is 'a fundamental, yet elusive aspect of public life'. This intersubjective intimacy can be thought of, I want to argue, as a particular mode of social hospitality.

Laurier and Philo (2006) explore this relation in their café work in great detail, analysing video footage of interactions between customers performing a welcome to other customers – letting them share space at a table, for example – or conversely giving an unwelcoming 'cold shoulder' that says 'Don't sit at my table' or 'Don't try to talk to me'. Their concern with what they call 'the architecture of our everyday intersubjectivity with others (known and unknown)' (2006, 195) is founded on their observation that 'the space of unacquainted persons in public space is heavily gestural and lightly conversational' (197) – though such space is intensely *communicative*. What gets communicated, I want to suggest, is the welcome, the feeling of belonging, gestured back and forth in moments of hospitality.

These kinds of mundane, taken-for-granted, fleeting, mobile performances are an important but often overlooked component of producing the 'hospitable city' (Bell, 2007).[3] Closer attention to the practices of performing 'friendliness', or 'non-threateningness', is therefore needed, to explore how hosting and guesting involve a 'welcome' given and received (or not), and how sociality in public spaces marks a place as 'warm' or 'cold', as 'friendly' or 'unfriendly' and so on (Morrill, Snow and White, 2005). Codes of behaviour between strangers, as well as between friends, family, colleagues, lovers, neighbours, and so on, therefore produce and reproduce intersubjective experiences of hospitableness (or not) for all participants.

Moreover, given that current mobilities are heavily technologically afforded and mediated (we might shorthand this as 'techno-mobility'), we must also ask about the roles of various non-humans in these performances: non-humans such as the built environment itself, and all the 'mobility devices' implicated in the rewiring of public and private highlighted above; as Sheller and Urry (2003: 113) note, 'the changing forms of physical and informational mobility that uproot bodies from place and information from space are key'. It is to these high-tech 'mobile hosts' that I now want to turn.

'Virtual' Hosting and Guesting

As noted earlier, the term 'mobile host' is used in discussions of mobile computing to describe a computational device which can remain networked while mobile, providing its user with an experience of 'seamless' connectivity, ignorant of the 'paths' and 'nodes' that are being switched between while 'on the move' (Perkins, Myles and Johnson, 1994). 'Hosting' in a computational context carries many connotations, in fact, such as the 'web host' that contains and enables access to content (web pages), making talk of non-human hostings more commonplace in computer culture. Moreover, the increasing convergence and intermodality of mobile information/entertainment/communications devices and personal computers is

3 Daniel Neyland makes an important intervention through his discussion of how immobility might be viewed as suspicious, given the normalization of mobility in cityspaces – 'immobile others' might be granted a less hospitable welcome when 'the "good city" is associated with the "mobile city"' (2006, 365).

radically transforming the experience of 'techno-mobility', even making it difficult to find suitable names to give to the gadgets and the practices they afford: the label 'mobile phone' or 'cell phone' is woefully inadequate, for example, to describe the multi-function hand-held device that can now combine voice communication, text messaging and email, global positioning data, Internet access, digital still and video camera, MP3 audio and video file player, calculator, personal organizer, and so on. As Manuel Castells (2005: 54) says, 'moving physically while keeping the networking connection to everything we do is a new realm of human adventure, one about which we know very little'. It is also a new realm to consider issues of hospitality, of hosting and guesting, as they are rewired by techno-mobilities.

Exploring the changing roles, meanings and experiences of what she refers to as 'mobile digital devices', Adriana de Souza e Silva (2006) describes a key transformation to the 'interfacing' capacity of such technologies: where the interface once mediated between the user and the machine – as in the graphic user interface that we see on a PC screen[4] – mobile digital devices mediate a 'social interface' between users, and also between users and the 'hybrid' (at once 'physical' and 'virtual') spaces they move through. As she defines this new space, 'hybrid spaces are mobile spaces, created by the constant movement of users who carry portable devices continuously connected to the Internet and to other users' (Souza e Silva 2006: 262).

Crucially, these new devices make it possible to be constantly connected both to networks such as the Internet and to other (human) users while moving through this hybrid space, thereby enfolding remote and proximal contexts and contacts: rather than disembedding the user, as personal computers have been thought to do when users enter cyberspace and 'leave the meat behind', users can now 'carry' cyberspace through physical space ('meatspace').[5] Moreover, emerging patterns of use suggest that 'mobile interfaces are used primarily inside social public spaces' (Souza e Silva 2006: 268) – yet rather then privatizing those spaces, these devices 'enfold' space, remixing public and private, physical and virtual, meatspace and cyberspace. Gaming applications illustrate this enfolding especially vividly, layering physical and virtual landscapes for players. Ultimately, for Souza e Silva, these new technologies 'recreate urban spaces as multi-user environments' (274).

Clearly, there are profound implications for the multiple users of these environments, therefore. In terms of thinking about hospitality, about mobile hostings and guestings, this new hybrid space changes the ways we think about host-spots and guest-spots, as both copresence and telepresence are enfolded and hybridized. Hosting and guesting are at once disembedded and re-embedded; like

---

4 The graphic user interface (GUI) performs its own version of hosting, of course, not only in assuming a level of computational competence in users, but also in offering various forms of assistance, handy tips and short-cuts, whether we require them or not (see Bell 2001 for a discussion).

5 'Meat' is a term used in certain cybercultural formations, especially those influenced by cyberpunk, to describe the physical, material, biological body: a body that stubbornly prevents us from uploading our consciousness and entering cyberspace fully, leaving the embodied realm of 'meatspace' behind (Bell et al., 2004).

the computational mobile host, which allows movement while securing connection to 'home' (via the protocol), reconfigured users in reconfigured spaces click between performances of host-ness and guest-ness as their devices seamlessly shift function. As with earlier transformations brought about by digital technologies, such as the shift from one-to-many to many-to-many broadcasting that characterizes the 'me media' aspects of the Internet and which subverts the 'old media' relationship between producer and consumer, this new hybrid space potentially rewrites forms and experiences of technologically mediated sociality, including relations of hospitality. While it is difficult to speculate about precisely how this reconfiguration of hosting and guesting will develop into particular new practices of hospitality, the mixing-up of proximity and telepresence in new hybrid spaces is suggestive of new forms of virtual hospitality between users who may or may not be 'physically' co-present and at once 'virtually' co-present. How are these hybrid spaces made into spaces of hospitality? Writing imaginatively at the dawn of the virtual age, Michael Benedikt (1991) suggested some principles of cyberspace design in order to limit the potential for disorientation and 'sim sickness'. Some of these key principles focused on social relations, such as the Principle of Commonality (a shared social reality) and the Principle of Personal Visibility (users should be able to 'see' who else is around them). While there is not a Principle of Hospitality in Benedikt's list, emerging codes and conventions such as 'netiquette' are part of this still emerging question: who is a guest and who is a host in cyberspace?

Other recent and forecast transformations in 'techno-mobility' offer us different takes on hosting and guesting. Some 'technological optimists', as Mimi Sheller (2004: 41) tags them, foresee ahead a future of bespoke datastreams and personalized messages as we move through urban environments crackling with ubiquitous computing or ambient intelligence. Anne Galloway (2004) surveys some recent experiments in ubiquitous computing, given the suitably sci-fi sounding acronym Ubicomp, providing a nuanced reading of the social and cultural implications of this developing set of technologies and practices. Ubicomp offers context-aware, 'invisible' or 'ambient' computing, through both location-aware mobile devices and devices embedded in particular environments. A cinematic depiction of one form of Ubicomp is provided in the science fiction thriller *Minority Report* (dir. Steven Spielberg, 2002), in a scene where the main character, John Anderton (Tom Cruise) rushes through a shopping mall. Retinal scanning technologies identify him, and virtual billboards address him personally, offering products tailored specifically to his profiled needs (The hidden surveillance function of such technologies plays out in this scene, since Anderton is on the run and attempting to elude capture; where once the space of the mall might have provided a cloak of invisibility, Ubicomp immediately and continually points up his location through its virtual sales patter.)

Current developments such as GPS (global positioning systems) and RFID (radio frequency identification) offer similar possibilities, providing locational data tagged to personal information (spending profiles mined from credit card purchases, or web browsing habits logged by cookies, for example). While wary of commercial applications such as these, Galloway notes how Ubicomp promises 'to create unique forms of habitable space and means of habitation' (2004, 389), discussing a series of experiments in producing what she calls 'amplified and annotated' city spaces.

A personalized tourist map sent to a PDA or mobile phone, for example, calibrates distances to the user's normal walking speeds and edits cartographic information, highlighting only sites of known interest to that individual user. A Swedish Ubicomp project, *Tejp*, aims to explore 'various possibilities for overlaying personal traces and information on public spaces', thereby transforming 'spectators into players and encourag[ing] playful ways to personalize territory in the public realm' (*Tejp* website, quoted in Galloway, 2004: 394). In one *Tejp* prototype, participants can deposit audio files, or tags, in public spaces, and other participants can hear those whispered messages as they pass by. The experience of public space is thus augmented by these audio traces, which produce a collective conversation about the city. Educators interested in 'blended learning' envisage similar systems with pædagogic applications, encouraging students to integrate mobile and Ubicomp technologies into active and engaging learning experiences on smart campuses (see for example: Kukulska-Hulme and Traxler, 2005).

Galloway approaches these innovations through a cultural studies lens, specifically via critical work on 'everyday life'. Questions she sees raised by Ubicomp when read through this lens – questions of interest to us here, too – include: 'How are space and time in the city negotiated by wireless technologies? What might constitute a sense of "place" in such scenarios? How might ubiquitous technologies map mobile experiences of everyday life?' (Galloway, 2004: 397). To which we might add further queries: how does Ubicomp reconfigure performances and experiences of hosting and guesting? Does Ubicomp also promise new forms of hospitable space and means of hospitality? The vision conjured, of ubiquitous hospitality, of hosting and guesting intermediated by mobile and ambient technologies, certainly challenges us to further rethink how hospitality can be performed and experienced in public spaces. As Benedikt (1991) prefigured, this raises a lot of complex questions regarding different forms of presence, different experiences of temporality and spatiality, different ways of understanding what it means to be 'together'. Ubicomp applications sometimes imagine rewiring public space as a personalized datascape, offering tailored welcomes and the promise of intermediated hosting (enabled through information). Some of the more 'intimate' emerging web spaces, such as blogs, may offer a glimpse into the reconfigurations of intersubjective hospitableness that Ubicomp might in the future more fully realize. As these technologies unfold, we are sure to witness further rewritings and rereadings of hospitality. Although somewhat different from past predictions of the delegation of routine hospitality to robot domestic servants, these developing technologies are bringing into being an equally radical transformation in the practices, relations and spaces of hospitality.

Conclusion

My focus in this chapter has been on both new and old mobilities – walking across a public square or surfing the Web from a portable digital device – seeking to understand how these are implicated in reshaping the host-guest relationship at the heart of conceptualizations of and experiences of hospitality, especially in cities and cityspaces. The host-work and guest-work that takes place in many forms across countless sites is surely a key index of the 'good city'. Analyses of the micro-practices currently taking place in sites of hospitality such as cafés or city squares, and discussions of future forms of hybrid, ambient or ubiquitous hospitality, reveal the endless ways that hospitality is given and received beyond the classical formulation of the host-at-home and the guest-visitor. New mobilities and new technologies further stretch and morph such a formulation, providing myriad new moments of hospitality. In particular, changing rhythms and temporalities of hospitality, the flickering moments, open up a reassessment of the ways in which hosting and guesting are articulated, performed, given and received.

References

Amin, A. (2006), 'The Good City', *Urban Studies*, **43**, 5–6, 1009–1023.

Banks, M. (2006), 'Moral Economy and Cultural Work', *Sociology*, **40**(3), 455–472.

Bell, D. (2001), *An Introduction to Cybercultures* (London: Routledge Books).

—— (2007), 'The Hospitable City', *Progress in Human Geography* **31**(1), 7–22.

Bell, D. and Binnie, J. (2005), 'What's Eating Manchester?, Gastro-culture and Urban Regeneration', *Architectural Design*, **75**(3), 78–85.

Bell, D., Loader, B., Pleace, N. and Schuler, D. (2004), *Cyberpsace: The Key Concepts* (London: Routledge Books).

Benedikt, M. (1991), 'Cyberspace: Some Proposals' in Benedikt (eds).

——, eds (1991), *Cyberspace: First Steps,* (Cambridge MA: MIT Press).

Binnie, J., Holloway, J., Millington, S. and Young, C., eds (2005), *Cosmopolitan Urbanism* (London: Routledge Books).

Bourdieu, P. (1984), *Distinction: a Social Critique of the Judgement of Taste* (London: Routledge Books).

Castells, M. (1996), *The Rise of the Network Society* (Oxford: Blackwell).

—— (2005) 'Space of Flows, Space of Places: Materials for a Theory of Urbanism in the Information Age' in Sanyal (eds).

Cochrane, A., Peck, J. and Tickell, A. (1996), 'Manchester Plays Games: Exploring the Local Politics of Globalisation', *Urban Studies*, **33**(8), 1319–1336.

Crang, P. (1994), 'It's Showtime: on the Workplace Geographies of Display in a Restaurant in Southeast England', *Environment and Planning D: Society and Space,* **12**(6), 675–704.

Derrida, J. (2001), *On Cosmopolitanism and Forgiveness* (London: Routledge Books).

Du Gay, P. (1996), *Consumption and Identity at Work* (London: Sage Publications).

Edensor, T. (2000), 'Staging Tourism: Tourists as Performers', *Annals of Tourism Research*, **27**(2), 322–344.

Florida, R. (2002), *The Rise of the Creative Class* (New York: Basic Books Books).

—— (2005), *Cities and the Creative Class* (New York: Routledge Books).

Galloway, A. (2004), 'Intimations of Everyday Life: Ubiquitous Computing and the City', *Cultural Studies*, **18**(2–3), 384–408.

Goffman, E. (1963), *Behavior in Public Places: Notes on the Social Organisation of Gatherings* (New York: Free Press).

Graham, S., ed. (2002), *The Cybercities Reader* (London: Routledge Books).

Gratton, C. and Henry, I., eds (2001), *Sport in the City: the Role of Sport in Economic and Social Regeneration* (London: Routledge Books).

Hochschild, A. (1983), *The Managed Heart: The Commercialisation of Human Feeling* (Berkeley, CA: University of California Press Books).

Knox, P. (2005), 'Creating Ordinary Places: Slow Cities in a Fast World', *Journal of Urban Design*, **10**(1), 1–11. [DOI: 10.1057/palgrave.udi.9000139]

Kukulska-Hulme, A. and Traxler, J., eds (2005), *Mobile Learning* (London: Routledge Books).

Lashley, C. (2000), 'Towards a Theoretical Understanding' in Lashley and Morrison (eds).

Lashley, C. and Morrison, A., eds (2000), In *Search of Hospitality: Theoretical Perspectives and Debates* (London: Butterworth-Heinemann Books).

Latham, A. (2003), 'Urbanity, Lifestyle and Making Sense of the New Urban Cultural Economy: Notes from Auckland, New Zealand', *Urban Studies*, **40**(9), 1699–1724.

Laurier, E. and Philo, C. (2006), 'Cold Shoulders and Napkins Handed: Gestures of Responsibility', *Transactions of the Institute of British Geographers*, **31**(2), 193–207.

Laurier, E., Whyte, A. and Buckner, K. (2001), 'An Ethnography of a Neighbourhood Café: Informality, Table Arrangements and Background Noise',, *Journal of Mundane Behavior (website)*. http://mundanebehavior.org/issues/v2n2/laurier.htm, (accessed: 23.08.06).

Lees, L., ed. (2004), *The Emancipatory City? Paradoxes and Possibilities* (London: Sage Publications).

McRobbie, A. (2002), 'Clubs to Companies: Notes on the Decline of Political Culture in Speeded-Up Creative Worlds', *Cultural Studies*, **16**(4), 516–531.

Miller, V. (2006), 'Intimacy and the Production of Public Space: the Intersubjective Construction of Enclaves', *Geographische Zeitschrift*, forthcoming.

Montgomery, J. (1995), 'Urban Vitality and the Culture of Cities', *Planning Practice and Research*, **10**(2), 101–109.

—— (1997), 'Café Culture and the City: the Role of Pavement Cafes in Urban Public Social Life', *Journal of Urban Design*, **2**(1), 83–102.

Morrill, C., Snow, D. and White, C., eds (2005), *Together Alone: Personal Relationships in Public Spaces* (Berkeley, CA: University of California Press Books).

Newman, H. (2000), 'Hospitality and Violence: Contradictions of a Southern City', *Urban Affairs Review*, **35**(4), 541–558.

Neyland, D. (2006), 'Moving Images: the Mobility and Immobility of "Kids Standing Still"', *Sociological Review*, **54**(2), 363–381.

Oldenburg, R. (1999), *The Great Good Place* (New York: Marlowe).

Peck, J. (2005), 'Struggling with the Creative Class', *International Journal of Urban and Regional Research*, **29**(4), 740–770.

Perkins, C., Myles, A. and Johnson, D. (1994), 'IMHP: a Mobile Host Protocol for the Internet' [website], http://citeseer.ist.psu.edu/cache/papers/cs/1963/ftp:zSzzSzftp.mpce.mq.edu.auzSzpubzSzeleczSzdistzSzmobilezSzimhpzSzjenc94zSzinet94.pdf/perkins94internet.pdf, (accessed: 20.08.06).
Pratt, A. (2002), *Creating a Regional Framework for the Cultural Sector* (London: DCMS).
Richards, G. (2000), 'The European Cultural Capital Event: Strategic Weapon in the Cultural Arms Race?', *International Journal of Cultural Policy*, **6**(2), 159–181.
Roberts, M. and Turner, C. (2005), 'Conflicts of Liveability in the 24-Hour City: Learning from 48 Hours in the Life of Soho', *Journal of Urban Design*, **10**(2), 171–193.
Roche, M. (2000), *Mega-events and Modernity: Olympics and Expos in the Growth of Global Culture* (London: Routledge Books).
Rogers, R. (1999), *Towards an Urban Renaissance: Final Report of the Urban Task Force* (London: E & FN Spon).
Ross, A. (2003), *No Collar: The Humane Workplace and its Hidden Costs* (New York: Basic Books Books).
Sanyal, B., ed. (2005), *Comparative Planning Cultures* (London: Routledge Books).
Sheller, M. (2004), 'Mobile Publics: Beyond the Network Perspective', *Environment and Planning D: Society and Space*, **22**(1), 39–52.
Sheller, M. and Urry, J. (2003) 'Mobile Transformations of "Public" and "Private" Life', *Theory, Culture & Society*, **20**(6), 107–125.
——, eds (2004), 'Places to Play, Places in Play' in Sheller and Urry (eds).
——, eds (2004), *Tourism Mobilities: Places to Play, Places in Play* (London: Routledge Books).
——, eds (2006), 'The New Mobilities Paradigm', *Environment and Planning A*, **38**(2), 207–226.
Shibli, S. and Gratton, C. (2001), 'The Economic Impact of Two Major Sporting Events in Two of the UK's "National Cities of Sport"' in Gratton and Henry (eds).
Short, J.R. (2000), *The New Urban Order* (Oxford: Blackwell Books).
Souza e Silva, A. (2006), 'From Cyber to Hybrid: Mobile Technologies as Interfaces of Hybrid Spaces', *Space and Culture*, **9**(3), 261–278.
Skeggs, B. (2005), 'The Making of Class and Gender through Visualizing Moral Subject Formation', *Sociology*, **39**(5), 965–982.
Stevens, Q. (2004), 'Urban Escapades: Play in Melbourne's Public Spaces' in Lees (eds).
Urry, J. (2000), *Sociology Beyond Societies: Mobilities for the Twenty-first Century* (London: Routledge Books).
Wittel, A. (2001), 'Towards a Network Sociality', *Theory, Culture & Society*, **18**(6), 31–50.

Chicks Pump Iron (Zacatecas, Mexico, 2006) by Elly Clarke

Chapter 3

Hospitality and Migrant Memory in Maxwell Street, Chicago[1]

Tim Cresswell

This chapter considers the ways in which past mobilities are made welcome (or, on the other hand, excluded) in the constitution of present places. It explores a particular place, Maxwell Street in Chicago, which has historically been the site of considerable migration both from abroad and internally. Maxwell Street is part of an area of Chicago which acted as a magnet to East Europeans, Jewish people, African-Americans and, more recently Hispanics. It was the site of the largest outdoor market in the United States until the 1990s, when the site was closed so that the nearby University of Illinois could build sports fields, parking lots and expensive apartment complexes. It was for this reason that a group of local activists sought to prevent the complete annihilation of this place of migrant memory by seeking to have it placed on the Register of Historic Places twice: first as a place associated with Jewish immigration and market stall activity and second as a site associated with African-American performance of the blues. Both applications failed. This chapter considers the arguments made for and against these applications in the light of theorizations of place as both a material structure and a palimpsest of mobile migrant practices. It considers the difficulty in institutionalizing sites of memory when those memories are of practices that were necessarily transitory and mobile and left no obvious mark on the material landscape. The chapter will thus relocate the discussion of hospitality in two principle ways. First hospitality is usually used to refer to contemporary migration and this chapter considers the question of whether it is possible to be hospitable to past mobilities. Second hospitality usually refers to reactions to the movement of people. This chapter considers the possibility of being hospitable to non-human objects.

In addition to hospitality then, this chapter considers the convoluted interconnections between place, mobility, memory, materiality and practice in the context of Maxwell Street. The chapter proceeds as follows. First I introduce Maxwell Street and its market. Following this I reflect on the interactions between place and mobility both theoretically and in the context of Maxwell Street. This includes accounts of the people and things at the market as well as discussions of

1 I would like to acknowledge the assistance of Lori Grove and Steve Balkin in the writing of this chapter. It would have been impossible without their leads and access to the archives of the Maxwell Street Historic Preservation Society. The research was conducted with assistance from a British Academy Small Grant.

sound, smell and garbage. The second half of the chapter reflects on contemporary (failed) attempts to preserve something of the milieu of the market in the face of redevelopment and gentrification. Towards the end of the chapter I consider how some objects are valued while others are not and consider the theme of hospitality in relation to objects and the past.

Maxwell Street

Maxwell Street and the area around it has played an important role in Chicago's history as a site of constant migration, from Europe and from areas further East and South in the United States. Some have referred to it as the 'Ellis Island of the Midwest' due to the fact that it was the first port of call for many immigrants to the city (Lawrence, 1999). In the 1920s, when over one-third of Chicago's population was foreign born, the area around the intersection of Maxwell Street and Halstead Street was home to immigrants from at least a dozen nations. Originally, these included German and Dutch populations fleeing persecution in Europe but by the 1880s the area became the home to a large Jewish population who quickly established an informal market consisting of peddlers' carts and open stalls selling everything from live chickens to tailored clothes. In 1912, the market was officially recognized as 'Maxwell Street Market' by the city council (Berkow, 1977; Eshel and Schatz, 2004). Beginning in the 1920s Maxwell Street experienced the arrival of African-Americans migrating into Chicago from the American South (Lemann, 1991). Over the next decades the area became an important centre of black culture in the United States and is credited as being the birth place of the urban electric blues (also known as the Chicago Blues) as blues musicians were able to plug electric guitars into the electrical outlets provided by the stalls set up for the market. By the early 1990s the area had become more strongly associated with Hispanic migrants. Indeed, the currently existing Maxwell Street Market (relocated to Canal Street) is notable for its excellent Mexican food. The market, in its heyday, was the largest outdoor market in the United States. It existed in one form or another around the intersection of Maxwell and Halstead until 1994 when the nearby University of Illinois, Chicago (UIC) decided to build new playing fields, parking lots and upmarket apartment buildings on the site of the old market. It was at this point that a group of Maxwell Street denizens sought to enter the area in the National Register of Historic Places in order to stop the university from demolishing the whole area. It was one of many radical and not so radical actions taken to counter the university-led gentrification process. Before considering these attempts, however, it is necessary to consider how place and mobility came together in this unique urban space.

Place, Mobility and Hospitality

How can places be hospitable to memories? Places are famously complicated things. First they are material things with definite shape and substance – buildings, roads, parks. Second they are meaningful – they evoke senses of place some of which are personal and some of which are shared. Third they are practiced – they are sites for action. People act out their lives in places and places have particular meanings, in part, because of the things people do in them. Materiality is possibly the most obvious aspect of place. Certain things in place can be preserved and protected while others are knocked down or allowed to decay. Materiality has the quality of obduracy and persistence. The meanings of place can also be preserved and protected but are more malleable and potentially ephemeral. Some meanings are collective and remain constant through time. But even the most public symbol has other meanings – counter-meanings – that contest the already established. Meanings, unlike materiality, are difficult to preserve in a pure form. They are easily contested and transformed. Practices are still more ephemeral – they do not remain after they happen. While their traces may be inscribed in the structure of things the actions themselves need to be repeated again and again to retain association with place. All three of these facets of place – materiality, meaning and practice – exist within social relations marked by power. The materiality of place is the result, more often than not, of planners, architects, investors, corporations and governments. It is only at the micro scale that each of us has the power to fashion concrete places. Meanings can be individual but insofar as they are shared they are legislated by powerful groups in society. The meaning of Stonehenge, for instance, was narrated by English Heritage and the UK Government in the 1980s in such a way that the alternative histories of travellers, hippies and others were marginalized and excluded (Cresswell, 1996; Hetherington, 2000). Finally practices are obviously legislated by law and custom. While forms of transgression and resistance will always occur the majority of our practices conform more or less to what is expected of us. It is at the level of practice, however, that we have the greatest freedom to create and transform places for ourselves.

Now consider mobility. Mobility has often been cast in opposition of place. Mobile practices have been said to undo place and threaten its established identities (Relph, 1976; Tuan, 1977). But recent scholarship has pointed towards how places are produced and reproduced through mobilities (Massey, 1997; Cresswell, 2002). The place-ballet of people's movements on a daily basis make up the particular rhythmic geographies that make places specific (Seamon, 1980). Places are sites where diverse people following diverse routes come together. But mobility, like place, is complicated (Cresswell, 2006). Mobility involves the material fact of movement – getting from A to B. But just as place is more than location so mobility is more than movement. Mobilities, like place, are made meaningful in a social context. The daily commute of the businessman in his BMW is very different from the movements of recently arrived immigrant domestic cleaners who are often dependent on public transport. Human mobilities are also embodied practices. We feel the pain in our joints as we walk through the city. We feel the jetlag and dehydration as we fly across the Atlantic. But it is not just people that move. As John Urry, has argued, the theorization of mobility should also include the mobility of objects and ideas as well as more imaginative and virtual mobilities (2000).

Following Jacques Derrida, questions of hospitality have tended to focus on hospitality towards moving people who enter a place and confront that place and the people who inhabit it with new challenges. How can a place (the nation-state, the city, the neighbourhood) accommodate the new arrivals and how much hospitality can be extended? How, on the other hand, might these mobilities be legislated and regulated? Questions of place and mobility are clearly central to the question of hospitality. In Maxwell Street migrants established new lives both at the market and in the nearby neighbourhoods. In many ways it was a very hospitable place. But it was not only the mobilities of immigrants and migrants that confronted the place. The market was a place of consumption – a place where *things* moved through along with shoppers and curious observers. Tourists even.

Place and Mobility in Maxwell Street

So what kinds of mobilities make up a place like Maxwell Street? First there are human mobilities. Maxwell Street is famously a site of immigration – the Ellis Island of the Midwest. It has been home to a dazzling succession of immigrant groups. It is a thoroughly migrant place. People also passed through this place every Sunday for over a century as they attended the market. On some days over 100,000 people would visit. While there, the bargain hunters would experience all kinds of mobile practices. Sellers would hawk their diverse wares through a number of elaborate performances. 'Pullers' would gesticulate to the passing pedestrians attempting to draw them into the stores that lined the streets. Musicians would perform the electric blues hoping for a donation.

> On Sundays additional hundreds appear from everywhere to offer goods obtained from fire sales, war surplus, manufacturers' odd lots, and the junk yards. The gypsy fortune-tellers, fake swamis, yodelling fish peddlers, and amateur entertainers come out. Buyers, bargainers, and sight-seers throng in, representing many nationalities and many races and more parts of the country. On Sundays, Maxwell Street is in its full glory.
>
> (Lloyd Wendt, *Sunday Tribune*, 'Business is always good on Maxwell Street', 9 October 1947, no page number)[2]

This open-air market was an immigrant space to be sure but it was also a theatrical event enacted repeatedly by sellers and buyers on a weekly basis. As the *Sunday Tribune* journalist observed (above) some of the visitors were 'sight-seers'; tourists who included the market on their itinerary of Chicago 'sights'. The fact that Maxwell Street was both a site of difference (a place obviously marked by the foreignness of the immigrant stall-holders and inhabitants) and a promiscuous place of objects for sale attracted tourists simply to watch and wonder. Maxwell Street was hospitable to these mobilities too. Maxwell Street was defined more by these mobile practices than by the material landscape that surrounded the market. Hardly a mention is made of the buildings that existed behind the stalls in over a century of newspaper coverage.

Materiality made its presence felt in a different way. Massive amounts of 'things' moved through the market. Observers were astounded by the range of things

---

2 The empirical data for this paper are drawn from a range of sources. They include the news clipping archive at the Chicago Historical Society Library and the personal archives of Maxwell Street activists Steve Balkin and Lori Grove. I am grateful for their help in this. Some news clippings do not have a page number marked.

they encountered at Maxwell Street just as much as the range of people and their practices.

> Shoes, clothing, fish, oranges, kettles, glassware, candy, jewelry, vegetables, crates of live poultry, hats, caps, pretzels, hot-dogs, ice cream cones, beads and beans, hardware and soft drinks, lipsticks and garlic are massed together in glorious ensemble of confusion.
> (*Chicago Daily News*, 'Maxwell Street Heart of Ghetto', 28 April 1928, page 14)

Again and again, when faced with the market, observers resorted to a poetry of abundance. Long lists of things became a standard market trope for journalists. All of these things passed through the market from uncertain places to the domestic spaces of the cities.

A Place of Excess – 'The dazzling excitement of a merry-go-round'

Place is often represented as a visual object. As Tuan reminds us however, places are far more than simply visual:

> We may say that deeply loved places are not necessarily visible, either to ourselves or to others. Places can be made visible by any number of means: rivalry or conflict with other places, visual prominence, and the evocative power of art, architecture, ceremonials and rites. Human places become vividly real through dramatization. Identity of place is achieved by dramatizing the aspirations, needs and functional rhythms of personal and group life (1977: 178).

Places are experienced, and experience involves all of the senses. We hear, smell, taste, feel and see them. This richness of place experience is evident in historical accounts of Maxwell Street. While observers were dazzled by the array of people and things at the market it was not the eyes that were most impressed with the kind of place the market was. Taste, sound and, particularly, smell were more prominent. The sound of the stall-holders' cries and various forms of music mixed with the smells of the various foods associated with the immigrants who ran the market.

> Barkers, spielers, pitchmen, and hucksters shout their wares while radios boom and customers haggle in a dozen languages. Merchandise drapes from awnings, spills over sidewalk stands and creaking pushcarts, litters the pavement and walks wherever the hawkers elect to take their stand. There is the sharp odor of garlic, sizzling redhots, spoiling fruit, aging cheese, and strong suspect smell of pickled fish. Everything blends like the dazzling excitement of a merry-go-round.
> (Lloyd Wendt, *Sunday Tribune*, 'Business is always good on Maxwell Street', 9 October 1947, no page number)

Often the area was compared to some place in the Middle East or North Africa. The market was referred to as a 'bazaar' – a reference which displaced Maxwell Street from the American Midwest to Baghdad or Marrakech. The combination of noise and smell, in particular, suggested to observing journalists something distinctly foreign. The descriptions of excess along with the references to places far away are similar to travel narratives for and by tourists. By visiting Maxwell Street these journalists were vicariously visiting a foreign destination marked both by foreign people and by 'exotic' smells and tastes. The 'bazaar' references displaced both things in the market and the observer.

Tourists of course, and particularly the kind of tourist referred to as a 'traveller', mixed their desire with disgust. As Bauman has put it: 'Waste is simultaneously divine and satanic. It is the midwife of all creation – and its most formidable obstacle. Waste is sublime: a unique blend of attraction and repulsion arousing an equally unique mixture of awe and fear' (2003: 22). Vision and smell combined in a particularly telling way in the repeated references to garbage and waste. Indeed, the history of representations of Maxwell Street is swirling with references to garbage, to waste, to noxious smells and cacophonous noise.

> Maxwell Street is ankle deep in dirt and refuse, the buildings that line it are dilapidated and tottering. Each passing breeze wafts a rich emulsion of the odors of poultry, crates, day-before-yesterday's spinach, fresh fish, garlic, cheese and sausage; and along with this emulsion it wafts particles of solid matter and deposits them impartially over the counters piled high with prunes, gum-drops, rye bread, dill pickles and other kosher delicacies.
> (*Chicago Daily News*, 'Maxwell Street Heart of Ghetto', 28 April 1928, page 14.)

Throughout this history of accounts of Maxwell Street we are presented with visions of excess. The people are excessive, the things are excessive, the sounds and smells mingle promiscuously. References to garbage, pollution and filth underline the status of the area as an immigrant space. References to dirt, as Mary Douglas has famously suggested, are references to 'matter out of place' (1966). Smell too, has often been unwelcome in modern spaces unless it is carefully controlled perfume (Corbin, 1986). The sense of smell has been relegated to the bottom of a hierarchy of senses in modern life (Miller, 1997). 'That there are bad sounds need not diminish the glory of hearing. That there are delightful fragrances has done little to elevate smell: traditionally, the best odor is not a good odor, but no odor at all' (Trotter, 2005: 38). Descriptions of bad odours are particularly prevalent in the extensive literature on urban reform from the nineteenth century (Trotter, 2005). Here, smell was used to underline the otherness of the urban poor of London or Paris. Similarly in these accounts of Maxwell Street garbage, smell and sound mingle in uncontrolled ways. These images of garbage, smell and sound are linked to reference to people such as the homeless while the consistent references to 'Fez', the 'Bazaar' or 'Baghdad' take the reader to other places where the senses may be more used to such a barrage of sensation. A bad smell can create anxiety because 'it comes back from outside, from elsewhere' (Trotter, 2005: 39).

What is clear from the snapshots of Maxwell Street stories above, is that mobility and place interacted in endlessly fascinating ways. This place was seen as unique and interesting because of the kinds of mobilities that passed through it and were enacted within it. It was seen as an immigrant place that served as an entry point for recent immigrants into American culture and society. It was a space of human drama every Sunday when shoppers and stall-holders performed their rituals on a regular basis. It was a place of overwhelming things being sold as well as things classified as garbage or waste. It was a place of all the senses. Materiality, meaning and practice intermingled in unique ways.

The question of overwhelming 'excess' would seem to be a key one for the theorization of hospitality. The excessive is that which needs to be rejected, regulated, excluded, removed. It is simultaneously fascinating, desirable and exotic. Excess is not possible without order. As Bauman has argued, waste is a by-product of order-making (2003). The more you are concerned with order, the more noticeable is that which exceeds it. In Bauman's formulation the order-making processes of modernity produce people-as-waste – the homeless, the refugees, the migrants. It is these very people that Derrida suggests are in need of an impossible hospitality – a welcoming without limits (2001). It is comparatively easy to extend hospitality to those who fit in and belong to a place. It is excess/waste that makes hospitality a challenge. And while Maxwell Street has been inhabited by migrants who have been the 'excess' of other places (Eastern Europe, the American South, Mexico, and so on) it has also been the site of excessive things – of literal waste. This too, has proved a challenge. Things are not inherently excessive as waste. They become waste as they reach particular moments in their biographies when they are not commodities (Kopytoff, 1986).

So what of Maxwell Street now? The rest of this chapter considers arguments about how a place like Maxwell Street proceeds into the future.

717 West Maxwell Street

Let us begin with 717 West Maxwell Street. The building whose address was 717 West Maxwell Street was built in 1883 as a two-storey building. A third storey and rear extension was added in 1909. The extension was designed by the architect David Saul Klafter, well known for several much bigger buildings in the city including the Insurance Centre on South Wells Street and West Van Buren Avenue and a large neo-classical office block at 415 West Aldine Avenue. We do not know much about what happened in 717 West Maxwell Street in the years between 1909 and the present day. We know that originally it did not have a storefront but that one was added. We know that the ground floor was used for a range of commercial purposes and that the upper floors were rented out for residential uses. In the past 717 West Maxwell Street has been the home of the Maxwell Street Meat Market, Frank's lamp shop and a fish market.

The practices that went on in 717 West Maxwell Street and meanings associated with it over nearly a hundred years are more or less invisible to us. Its materiality, however, continues to matter. Klafter's addition to the building included several features that made it architecturally interesting. Nearly a century later an archaeological survey of the building noted the following features:

> On the west side of the façade there is a set of pilasters that extend from the sign frieze to the lintel level of the third story. The pilasters frame a triple window at the third story. There is also a single rectangular window on the east side of the façade. That window opening is now filled in with brick. Both the single rectangular window opening and triple window opening on the second story are also now filled in with brick. Above the

> third story window openings, at what would be the lintel line, there is a limestone belt course with moulded coping. Above this is a frieze of patterned brickwork. Above this is a pediment with limestone coping over the bay framed by the pilasters. At the flat east side of the façade, the limestone coping of the pediment extends to a belt course separating the frieze above the third story from the parapet above it. The pilasters are square with simple limestone bases and capitals, on and below the capitals there are carved limestone foliage ornaments.
>
> (Report of a Phase I Archaeological Survey of the Maxwell Street Area in Cook County Illinois (Prepared by Archaeological Research Inc.) 1994, page 43)

This description of 717 West Maxwell Street formed part of a concerted attempt to save the area from demolition – to make it seem worthy of persisting. Activists formed the Maxwell Street Preservation Society in order to have the area placed on the Register of Historic Places – an act that would have made it considerably harder for the university to demolish the area. In order to submit an application, a formal archaeological survey had to be carried out in order to establish which of the buildings in the area would be worthy of preservation – worthy of being designated, in National Parks Service jargon, a 'contributing object'.

As it turned out the front of 717 West Maxwell Street (at least) was 'preserved'. In an attempt to pacify those objecting to the destruction of the Market area Mayor Daley instructed the university to preserve 13 façades and to relocate them in a single block to act as a symbol of the area's past. Consequently, 717 West Maxwell Street is now on the south side of Maxwell Street. The façade to its left belonged to 727 West Maxwell Street and the one on the right is from 1245 South Halstead Street. Over the street is the façade of 722 West Roosevelt Street. Behind these facades are restaurants, coffee shops and a multi-story parking lot.

Fixity, Flow and the Past

The Maxwell Street Preservation Society sought to have the Maxwell Street area placed on the National Register of Historic Places twice, in 1994 and 2001. In the first instance they highlighted its history as a marketplace and its association with, particularly, Jewish immigration. In the second instance they emphasized the area's history as an entertainment space – as the birthplace of the blues. Both applications linked processes of migration to more local mobile practices of market selling and music. In both instances they failed. These failures were based on understandings of what gives a place 'integrity'. To the National Park Service a place has to have integrity to be deemed worthy of preservation. This notion of integrity, if Maxwell Street is anything to go by, is based largely on the materiality of place, and its capacity to resist change. The applications for Historic Place status failed because the National Parks Service believed that the material structure was not sufficiently similar to the material structure of the area before 1944 as the material structure was so transformed, they argued, it could no longer be associated with the activities that took place there.

Putting Maxwell Street on the National Register of Historic Places needed approval by, first, the State of Illinois (through the Illinois Historic Preservation

Agency (IHPA)), and finally the National Parks Service in Washington DC. In between, the Chicago City Landmarks Division was asked to agree or disagree with the IHPA assessment. The submission involved completing a lengthy archaeological survey of the area and an application form that were sent to the IHPA. Following a unanimous decision the first Maxwell Street bid was approved at this stage. The IHPA then prepared a 52-page document including photographs and details of specific properties in the area which was sent on to the National Parks Service for their decision along with a letter from the Illinois Historic Preservation Officer summarizing the opinion. Before the documentation could be sent to the National Parks Service however a copy was sent to the Chicago City Landmarks Division of the Department of Planning and Development and to Chicago Mayor Daley's office.

The report stated that the significance of the district lay in its 'ethnic and commercial history' between 1870 and 1944 and was particularly remarkable for its associations with a succession of immigrant groups. Key to the IHPA decision was that the area retained 'sufficient integrity of location, setting, feeling, association, and materials from the period of significance.' The memorandum that accompanied the report put great emphasis on the dynamism and vitality of the market.

> The buildings, often nondescript, which contained the storefronts and the apartments of their proprietors, and the streetscapes crowded with pushcarts and stands were the physical context in which the bustling commerce and acculturation took place. The historical significance of the area lies *not in the occurrence of particular events of note within its confines, but in the vitality that took place from day to day in the area.*
> (Memorandum from the IHPA to Chicago Landmarks Division, 9 March 1994, page 8)

The principle case for historic place status, then, was its intangible quality of everydayness. The very kinds of things that were evident in over a century's worth of newspaper accounts. Maxwell Street was a *fluid* place that was experienced in a myriad of ways.

The problem for Maxwell Street's supporters was that this intangible flow of experience, people and things did not fit with the notion of integrity of place enshrined in the National Parks Service criteria for what counts as an historic place. The first rejection came from the Chicago Landmarks Commission which argued strongly against the area's inclusion on the Register. They made a distinction between the *activities* that take place in an area and the *material structure* of the *place* itself. In a letter explaining their decision, Peter Bynoe of the Chicago Landmarks Commission wrote that it was their view that the area certainly had 'rich historical associations' but did not have 'sufficient integrity' to convey this history as much of the material landscape no longer existed. He asserted that the remaining buildings were insufficient to convey the historical significance of the area. Many of the buildings that did remain had been altered in a way that undermined their ability to convey the history of the area. Most tellingly perhaps the letter remarked on the nature of the events that had taken place in the nominated area.

> The street activity of the Maxwell Street Market is its primary historical association. Given the makeshift, transitory nature of this activity, the buildings alone cannot convey

> the historical feeling of the market. The National Register assists in the preservation of buildings, not such fluid activity as was the historical essence of Maxwell Street.
>
> (Letter from Peter Bynoe (Commission on Chicago Landmarks) to Ms Swallow (IHPA), 6 May 1994)

The Landmarks Division then made the case that only the material aspects of a place can be preserved and that *activities* are too 'fluid' to be preserved. This is despite the fact that the National Parks Service identifies seven aspects of integrity that might make a place preservable. One of these aspects is *association* whereby a place 'retains association if it is a place where the event or activity occurred and is sufficiently intact to convey that relationship to an observer.'

The Fluidity of Place

The issue of the fluid nature of the events that took place at Maxwell Street was one which arose throughout the process of applying for Historic Place Status. It was obviously on the minds of some of the people who wrote letters in support of the application. William Garfield, a retired planner and alumnus of UIC wrote that:

> Regarding the integrity issue, one must look beyond the loss of structures into the fabric of the function of the market. Buildings are not what makes Maxwell Street historic; street vendors, musicians, and ambience did and does. The complex bringing together of persons from all walks of life – poor, rich, black, white, young, old, Hispanic, Jewish, eccentric, intellectual – is what makes Maxwell Street historic. The continuous use of this same place (…) as a poor person's business incubator for 125 years is what makes Maxwell Street historic.
>
> (Letter from William Garfield to Ms Swallow (IHPA), 24 May 1994)

Despite the unanimous verdict of the IHPA that the District deserved nomination to the National Register, the State Historic Preservation Officer (who was responsible for forwarding the nomination to Washington D.C.) came to the conclusion that the District did not merit Landmark status due to the lack of *integrity* of setting, design and materials. According to him there simply was not enough left of the material landscape that had existed in the 'period of significance' (1880–1944). His letter focussed on the extensive renovations, that had meant that the buildings were no longer like they were, as well as the demolition of large numbers of buildings within the district. Noting the high level of popular support for the District's nomination evidenced by many supporting letters he wrote that he did 'not believe that the integrity issue has been considered or given sufficient weight by most people' (Letter from William Wheeler (State Historic Preservation Officer) to Beth Boland (National Park Service) 16 November 1994).

Carol Shull of the National Park Service wrote to Lori Grove in December 1994 informing her that: 'Due to the irretrievable loss of historic integrity, the Maxwell Street Market Historic District does not meet the National Register Criteria for Evaluation and thus is not eligible for listing in the National Register' (Letter from Carol Shull of National Park Service to Lori Grove, 12 December 1994). The character of the district, she continued 'has been destroyed over time by demolition, neglect and alteration. The district as a whole no longer retains the requisite qualities of design, materials, workmanship and setting to qualify for listing in the National Register.' The second nomination in 2001 failed for the same reasons.

A Material Fixation

So the members of the newly formed Maxwell Street Historic Preservation Society failed in their attempt to block the development of the University of Illinois 'University Village' which has now taken over the area. They failed because the material landscape of the area had changed. Despite this failure the 44-page list of buildings in the area complete with accompanying text drawn from an *Archaeological and Historical Evaluation of the Maxwell Street Area* had distinguished between 'contributing objects' and 'non-contributing objects' on a building by building basis. In each case the material structure of the building is assessed as to whether or not it would contribute to the 'integrity' of the place.

In the case of 717 West Maxwell Street the report details aspects of the material structure of the building such as the 'thin limestone belt course beneath the second story windows, and a band of similar limestone squares beneath this, extending horizontally across the sign frieze area.' Enough of these details were unmodified, the report states, 'the building has good integrity and would likely be deemed as a contributing feature to a potential historic district.' [3] 1310-1316 South Halstead Street, however, received a different assessment. Again the material structure is described in great detail but most of the focus is on transformation to the original building such as the removal of the top two stories, alterations to the storefront and the cladding of the building in metal siding. 'There is no evidence that an intact historic façade exists beneath the siding', the report asserts, and then, damningly, '[A]s the building has no historic integrity, it would likely be deemed as a non-contributing feature to a potential historic district.'[4]

In total the report features 44 'contributing features' and 10 'non-contributing features'. Each assessment is based on the degree to which the material fabric of the place is like it was in the 'period of significance'. Clearly 717 West Maxwell Street, despite the changes made to it over the years, was considered to possess 'integrity' while the material transformations of 1310-1316 South Halstead Street meant that it was 'non-contributing'. Somewhere between them is a point at which the degree of change switches from acceptable to not acceptable. Despite the fact that all places, and all material objects, are fluid and changeable, historical significance is attached in this reasoning to objects that are perceived to have stayed the same. Their 'persistence' became their most important characteristic. Mobility in time and space is conceptualized as the enemy of integrity.

So we can see how activists and proponents of preservation in Maxwell Street emphasized the vitality of a living landscape in their efforts to get some kind of

3 Report of a Phase I Archaeological Survey of the Maxwell Street Area in Cook County Illinois (Prepared by Archaeological Research Inc.) 1994, page 43.

4 Report of a Phase I Archaeological Survey of the Maxwell Street Area in Cook County Illinois (Prepared by Archaeological Research Inc.) 1994, page 52.

protection for the area while the officials responsible for making decisions emphasized unchanging materiality. But while relatively unchanging materiality was key to the preservation of a number of buildings and facades in the Maxwell Street area it would be wrong to suggest that materiality *itself* required preserving. Maxwell Street, as we have seen, was positively overflowing with things – some of them commoditiees and some of them garbage. The question is: *How do some things persist while others are discarded*? Compare, for instance, the record of 'contributing objects' which included 717 West Maxwell Street with another record of objects – a record compiled by members of the Maxwell Street Preservation Society. While the University of Illinois was busy knocking down the area around Maxwell Street, these Maxwell Street supporters were busy collecting things and storing them as records of what Maxwell Street once was. The record is titled 'Maxwell Street Archives and Artefact List' and it is arranged according to the address where the material originated and the place it is currently stored. These places of storage include members' basements, garages, gardens and the rear yard of a church. The list includes the following from 717 West Maxwell Street:

> French-fry maker (manufacturer: Bloomfield Industries, Chicago)
> (Collected by MM.)
> Document: "Order in the Circuit Court of Cook County, Illinois," no. EOL50734 dated 10/1/2002, for Zafar Sheikh
> One fragment of awning canvas, green, apr ox. 9"x7-1/2"
> (Collected by LG, 02.)
> There is also a record of what was collected from the ill-fated 1310-1316 South Halstead:
> Building Inspection Certificate for 1314-1/2 (S. Halsted), dated 1964
> Colt 45 plastic and paper sign with 3-D bottle, 15"x10"
> 'Only Remy,' tin sign, 10"x17"
> 1 Marlboro Race Car Illuminated Electric Sign, 31x 18' x 2'
> Misc. liquor store signs, small
> (Collected by SB, JW, MM, 2002)
> 4 part sign reading: 'KISER Rip Jew Town: '
> (Rip is thought to be abbreviation for Rest in Peace.)
> 'Polish Sausage Pork Chops Hamburger Hot Dog Fries,' 8' x 20"
> 'Free Fries with Sandwich,' 8' x 18"
> (Collected by SB, JW, MM, 2002.)

The eleven-page single-spaced list of items includes, bits of buildings (tiles, frames, doors, ceiling pieces, and so on), signs, paper bags, bits of wood, ashtrays, books, papers, excavated soil samples, a lamp shaped like a bear, shirts, a sewing machine, records, pushcarts and six tailor shop male manikins. The list of objects reveals the interests of some of the people collecting this material not simply in the buildings but in the material evidence of the life that went on in and around them. Though these were the things that made Maxwell Street what it was, these items are all undoubtedly non-contributing objects in the terminology of the National Parks Service, and therefore irrelevant to the conservationists' application for historical protection

Rubbish and the Mobility of Things

Many of the arguments that swirled around Maxwell Street were arguments about the values of things. Briefly put, discussion centred on whether certain objects in the Maxwell Street area, and the area itself deserved to persist or be discarded. Michael Thompson's book, *Rubbish Theory*, is helpful in this regard (1979). In this book Thompson outlines the importance of the category of 'rubbish' to the construction, maintenance and transformation of 'value'. Rubbish enables the assignation of value to objects that are not rubbish. Objects, in Thompson's theory, can be assigned the value of 'transient' or 'durable'. Transient objects decrease in value over time while durable objects either maintain or increase their value over time. The ways in which objects are categorized as durable or transient shapes the meanings assigned to these objects and the human relationships to and around them. Thus, an old Vauxhall car might be lent to teenage children while another old car might be labelled vintage and preserved in a collection.

> Signs of aging and use can contribute to increased or auratic value. In the same sense the shiny new plastic cup appears to us as imminent rubbish; disposability makes transient value strikingly visible. The fact of malleability and transformation in value is evidence that objects are not locked into categories because of their material categories. It is *how* their materiality is apprehended and used that is key to value and transformation (Hawkins, 2006: 78).

The key liminal category between the durable world and the transient world is 'rubbish'. Rubbish provides a categorical space where transience can be converted into durability. Rubbish denotes objects that have the possibility of being re-valued as durable – as valuable. One of Thompson's examples is run down inner-city houses that are gentrified and re-valued as 'desirable'. The assignment of things to the category of durable or transient is performed by those in control of time, space and knowledge who thus ensure 'that their own objects are always durable and that those of others are always transient' (Thompson, 1979: 9).

In Thompson's formulation, objects travel though a series of regimes of value – they have careers. 'No single game exhausts their function; no single description exhausts the uses to which their properties might appropriately or inappropriately lend themselves. Indeed, objects don't simply occupy a realm of objecthood over against the human: they translate human interests, carry and transform desires and strategies' (Frowe, 2003: 36). Kopytoff has made a similar argument about things having 'biographies' during which they enter and leave the category of 'commodity' (1986). In Appadurai's terms, they have 'social lives' (1986). In this sense the mobile objects of Maxwell Street – from old tin cans to 717 West Maxwell Street – are fluid concretizations of the relations between the human and the non-human worlds. These concretizations are, of course, contested. The official survey of relatively fixed, relatively stable, 'contributing features' includes objects valued for their ability to represent a version of the past. The objects on the activists' 'Maxwell Street Archives and Artefact List' represent a different relation between the human and non-human worlds – one that is interested in fluidity and the quotidian.

Conclusion

So how then, does this relate back to hospitality? In his essay *On Cosmpolitanism and Forgiveness* Derrida considers the notion of hospitality in relation to the idea of a 'city of refuge' – a place where all will be accepted regardless of point of origin – a locus of universal hospitality. He points towards the tension between the notion of hospitality as a universal right – in Kant's terms 'hospitality without limit' – and the actual granting of hospitality as a form of sovereign power – as, for instance, the granting of a right to residence. It is the gap between hospitality as a universal and unconditional right of cosmopolitanism and hospitality as something which is granted that has proven problematic within Western liberal societies trying to come to terms with issues of immigration and asylum (Derrida, 2001). Here I have considered a migrant city space – a space that has acted as just such a 'city of refuge' for over a century. The space was hospitable to both migrant people and mobile things. It was a place where people and things that were in effect the 'waste' of other places could be welcomed and, occasionally, re-valued (Bauman, 2003). The process of deciding whether or not such a space was worthy of preservation, I would argue, was an exercise in granting hospitality. The failure of the two applications for Historic Place status was a denial of hospitality. It denied a refuge in memory for both the immigrant groups that inhabited the space and the things that they brought through it. While certain kinds of fixity and permanence were declared worthy and possessing 'integrity', fluidity and practice were marginalized and excluded – deemed lacking in integrity – unworthy. Similarly the process of granting something durability rather than declaring it 'transient' enacted a particular kind of exclusion.

However hard the Preservationists tried to make an argument based on fluid practices and invisible senses of place, their arguments were trumped by the undeniable material transformations that had occurred in Maxwell Street over the last 50 years. Obdurate materiality is simply reified – in the end it is all that counts. This story ends with irony. The area that now exists where the market used to take place is called University Village. An invitation to an open day at the new 'oversize' apartments featured a picture of a market trader with an artist's impression of the new developments superimposed on top with the tag line 'Yesterday's Heritage… Tomorrow's Treasure'. In August 2000 the Chicago Tribune found new village residents Al and Mary Geiser admiring their new home. When asked what drew them to the village they replied that 'We liked the multi-ethnic feeling and the authenticity of the neighbourhood – we're intrigued by the history of Maxwell Street'. Mayor Daley, in an effort to pacify the preservationists agreed that the area had significant historical associations that needed to be preserved. He responded by protecting eight storefronts in the area and moved the facades of 13 buildings and attached them on to new buildings, including a large multi-storey parking lot. These facades include 717 West Maxwell Street. Here Maxwell Street is literally reduced to the material structure of facades that can be moved and relocated at will in order to produce a simulation of a notion of Maxwell Street at the heart of the university-owned development of shiny new apartments, sports fields and parking lots. Here the material signifiers of an earlier place are removed completely from the accompanying practices and performances that made Maxwell Street the place it was. And behind these facades of history are parked the cars of the owners of the new apartments that surround the area. Perhaps so they can feel rooted in an evocative past of street traders and blues musicians – a past deemed too fluid to merit the designation of historic place but strong enough to sell townhouses and condos.

References

Appadurai, A., ed. (1986), *The Social Life of Things: Commodities in Cultural Perspective* (Cambridge: Cambridge University Press).

Barnes, T. and Gregory, D., eds (1997), *Reading Human Geography* (London: Arnold).

Bauman, Z. (2003), *Wasted Lives: Modernity and its Outcasts* (Oxford: Polity Press).

Berkow, I. (1977), *Maxwell Street: Survival in a Bazaar* (Garden City, N.Y.: Doubleday Publishing).

Buttimer, A. and Seamon, D., eds (1980), *The Human Experience of Space and Place* (London: Croom-Helm Books).

Cohen, W. and Johnson, R., eds (2005), *Filth: Dirt, Disgust, and Modern Life* (Minneapolis: University of Minnesota Press).

Corbin, A. (1986), *The Foul and the Fragrant: Odor and the French Social Imagination* (Cambridge, MA: Harvard University Press).

Cresswell, T. (1996), *Place/Out of Place: Geography, Ideology and Transgression* (Minneapolis, MN: University of Minnesota Press).

—— (2002), 'Theorising Place' in Cresswell and Verstrate (eds).

—— (2006), *On the Move: Mobility in the Modern Western World* (New York: Routledge Books).

Cresswell, T. and Verstraete, G. (2002), *Mobilizing Place, Placing Mobility: The Politics of Representation in a Globalized World* (Amsterdam: Editions Rodopi Books).

Derrida, J. (2001), *On Cosmopolitanism and Forgiveness* (London and New York: Routledge Books).

Douglas, M. (1966), *Purity and Danger: An Analysis of Concepts of Pollution and Taboo* (New York: Praeger Publishing).

Eshel, S. and Schatz, R. (2004), *Jewish Maxwell Street Stories* (Charleston, SC: Arcadia Books).

Frowe, J. (2003), 'Invidious Distinction: Waste, Difference and Classy Stuff' in Hawkins and Muecke (eds).

Hawkins, G. (2006), *The Ethics of Waste: How We Relate to Rubbish* (Lanham, MD: Rowman & Littlefield Publishers).

Hawkins, G. and Muecke, S., eds (2003), *Culture and Waste: The Creation and Destruction of Value* (Lanham, MD: Rowman & Littlefield Publishers).

Hetherington, K. (2000), *New Age Travellers: Vanloads of Uproarious Humanity* (London: Cassell Books).

Kopytoff, I. (1986), 'The Cultural Biography of Things: Commoditization as Process' in Appadurai (eds).

Lawrence, C. (1999), 'City Dealing to Preserve Maxwell Street', *Chicago Sun-Times*, 12 March 1999, 6.

Lemann, N. (1991), *The Promised Land: The Great Black Migration and How it Changed America* (New York: A. A. Knopf).

Massey, D. (1997), 'A Global Sense of Place' in Barnes and Gregory (eds).

Miller, W.I. (1997), *The Anatomy of Disgust* (Cambridge, MA: Harvard University Press).

Relph, E. (1976), *Place and Placelessness* (London: Pion).

Seamon, D. (1980), 'Body-Subject, Time-Space Routines, and Place-Ballets' in Buttimer and Seamon (eds).

Thompson, M. (1979), *Rubbish Theory: The Creation and Destruction of Value* (Oxford and New York: Oxford University Press).

Trotter, D. (2005), 'The New Historicism and the Psychopathology of Everyday Modern Life' in Cohen and Johnson (eds).

Tuan, Y.-F. (1977), *Space and Place: The Perspective of Experience* (Minneapolis, MN: University of Minnesota Press).

Urry, J. (2000), *Sociology Beyond Societies: Mobilities for the Twenty-First Century* (London and New York: Routledge Books).

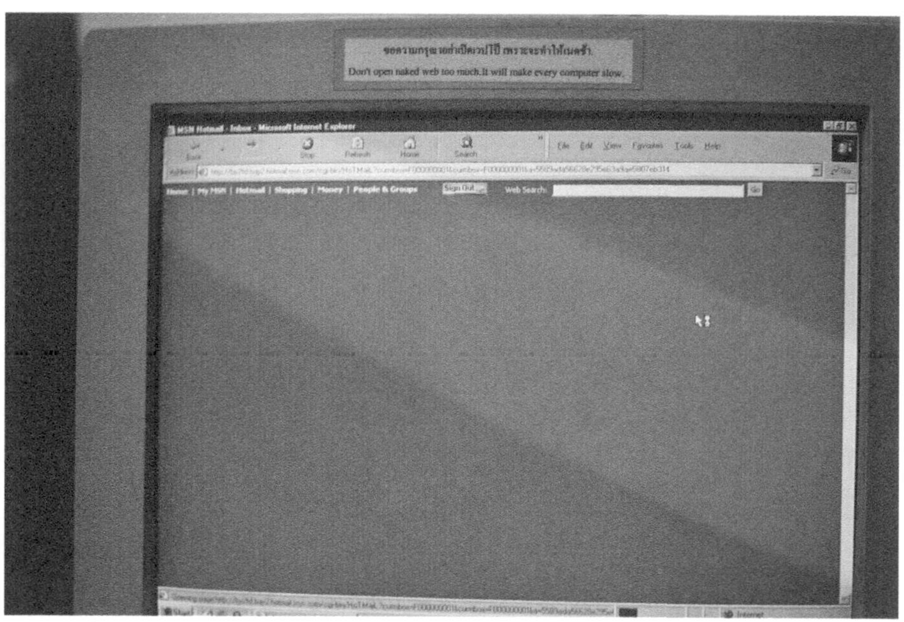

Naked Web **(Chiang Mai, Thailand, 2004) by Elly Clarke**

Chapter 4

Cosmopolitans on the Couch: Mobile Hospitality and the Internet[1]

Jennie Germann Molz

On 1 May 2001, Ramon Stoppelenburg left his home in Zwolle in the Netherlands to travel around the world for a few years. In many ways, the 24-year-old Dutch student was like thousands of other university students around Europe who take a year off to travel the world, usually on a shoestring budget. But, unlike other travellers, Ramon embarked on his trip without *any* money. In fact, Ramon planned to circumnavigate the globe on less than a shoestring: he planned to do it for free by hitchhiking and by staying with the hundreds of hosts who had invited him via his website *let-me-stay-for-a-day.com*. Months before his departure date, Ramon launched the website to allow people from around the world to submit invitations to host Ramon for a night. The 'deal', according to Ramon, was that in exchange for giving him a place to sleep for the night, a meal and Internet access, hosts would find their towns, their homes and possibly their cooking skills documented in the daily reports Ramon posted on his increasingly popular website.

According to newspaper coverage of the site, Ramon had become the world's most famous homeless guy – a Dutch freeloader for whom 'going Dutch' meant not paying at all. Ramon's website chronicled his ongoing journey in minute detail, allowing an ever-growing audience to follow virtually as Ramon travelled physically to the homes of over 500 hosts in a dozen countries during a two-year period. The discussion forum he hosted on the website also became an active hub where people could interact with Ramon and with each other. Here, past hosts and friends vouched for Ramon's charm and sincerity, future hosts expressed their eagerness to meet him and strangers from around the world debated, criticized and praised Ramon's project and his performance as a perpetual guest.

In addition to Ramon's *let-me-stay-for-a-day.com* project, an increasing number of hospitality organizations have appeared online to help travellers find a place to crash for a night or two. This chapter discusses several of these sites, including The *HospitalityClub.org.*, *Hospitality Exchange* at hospex.net, *Servas International*,

1 I am grateful to the participants of the Mobilizing Hospitality workshop for their engaging discussion of these ideas, to Sarah Gibson for her useful suggestions, to Alan Metcalfe and Jyri Engeström for pointing me in the direction of useful texts, and to Adi Kuntsman for her comments on this paper. Funding was provided for this research through an Economic and Social Research Council Postdoctoral Fellowship (Award PTA-026-27-0623).

GlobalFreeloaders.com and *CouchSurfing.com*.[2] These websites operate primarily as 'members only' communities that employ databases of thousands of members to connect hosts and guests online and face-to-face. Through the websites, members can search the database to look for a host in a particular destination, use profiles to see if what the host has to offer matches what the traveller needs, submit to various security measures intended to maximize the safety of these encounters between strangers and participate in active member discussion boards. Though a few of these sites charge a small membership fee, they are all not-for-profit organizations guided largely by the belief that world travel, interpersonal interactions between people from diverse cultures and the generosity of hospitality can spread tolerance, friendship and world peace at a grassroots level. These sites raise several challenging questions about hospitality as a social relation between strangers, about the links between mobility, hospitality and the Internet and about the way a 'global community' is forged, idealized and bounded through the prism of hospitality. In addition to coordinating the logistics of hospitality, these websites act as discursive sites where the moral, ethical and practical aspects of hospitality are contested and controlled.

In this chapter, I discuss several ways in which the Internet, both as a communication medium between geographically dispersed strangers and as a technological fantasy, is used to maintain the propriety of the hospitality relationship, to manage the anxieties inherent in moments of hospitality and to police the boundaries of the hospitable traveller-host community. First, I examine the way hospitality is understood as a *reciprocal* arrangement between hosts and guests. Like Kant, whose notions of hospitality Derrida explicitly critiques, these hospitality websites emphasize hospitality as a reciprocal exchange. Setting aside for the moment Derrida's critique of contractual formulations of hospitality (a critique I will return to later), I look at how these websites try to maintain the equilibrium of give-and-take in the hospitable encounter. Second, I consider how the *reputation* systems on these websites act as a kind of surveillance mechanism that monitors this reciprocity between hosts and guests, but that also secures the face-to-face meeting between strangers and controls the boundaries of the hospitality community. In both of these discussions, we can see how these websites also work to arrange the internal paradoxes that characterize definitions of hospitality. Drawing on Benveniste's (1973) etymology of the term 'hospitality', Derrida (2000) highlights its several contradictory meanings, such as 'host' *and* 'guest', 'guest' *and* 'parasite', or 'guest' *and* 'enemy'. Such indeterminacies give rise to certain anxieties surrounding hospitality, namely the risk that the guest may actually become a parasite or an enemy; a risk that the websites manage through reciprocity and reputation. Finally, I address the paradox of 'global community' as it is produced and performed in these hospitality communities. Here, I consider the way a cosmopolitan disposition of open-ness toward difference actually serves to delimit a bounded community of 'like-minded' but diverse individuals.

---

2 Apologies to readers expecting a Freudian account of mobile hospitality on the Internet; the chapter title employs 'the couch' not as a metaphor of psychoanalysis, but rather more literally in reference to website names such as *CouchSurfing.com*.

Reciprocity

The hospitality websites in this study govern interactions between strangers within an economy of hospitality that is negotiated in terms of *reciprocity*. In his analysis of the root terms of hospitality – *hospes*, *hostis* and *potis* – Benveniste argues that hospitality essentially denotes a reciprocal exchange. According to Benveniste (1973), the root *hostis* appeared in various Latin phrases expressing a sense of reciprocity: 'repay a kindness', 'I promise you a reciprocal service, as you deserve', and 'compensation of a benefit', with related terms referring to measurement and equalization (76). Benveniste explains that *hostis* signifies 'he [sic] who stands in a compensatory relationship… founded on the idea that a man [sic] is bounded to another (*hostis* always involves the notion of reciprocity) by the obligation to compensate a gift or service from which he has benefited' (77). In fact, according to such definitions of hospitality, it is precisely this *reciprocal* exchange that binds people in social solidarity. The websites I look at see hospitality similarly as a social pact, one that ensures not only the proper relationship between host and guest in the moment of the hospitality encounter itself, but also as a contract that extends a binding moral code across the whole community.

However, the symmetry of the hospitality encounter is always fragile. Because there is always the possibility of an excess, a lack or a slippage between give and take, the threat of imbalance – the prospect that the guest will take advantage of the host's generosity by taking too much or by *only* taking – must be constantly managed. In other words, guests must not be allowed to turn into *freeloaders*, imposing parasitically on other people's generosity without sharing in the cost or responsibility. For example, Gibson (2003) describes how unwelcome strangers, such as asylum seekers, are often portrayed in the media as parasites who take advantage of the nation's hospitality or welfare system without giving anything back. The websites serve to arbitrate this slippery equation, so it is somewhat confusing to find that in many cases these travellers happily refer to themselves as 'freeloaders' on the websites. Much of the media coverage of Ramon's journey refers to him in almost celebratory tones as the 'world's biggest freeloader'. And one of the websites, *GlobalFreeloading.com*, uses the word in its title and refers to its members as 'freeloaders'. Of course, these appropriations of the term 'freeloader' are purposefully ironic, foregrounding the anxiety of freeloading while at the same time making it impossible by strictly regulating the reciprocity between hosts and guests. By ironically appropriating terms like 'freeloader', these hospitality websites acknowledge the anxieties around parasitism while simultaneously distancing themselves from the highly politicized arena of immigration and 'national' hospitality.

Derrida (2000) points out that one of the uncertainties of hospitality is 'the general problematic of relationships between parasitism and hospitality' (59). He asks: 'How can we distinguish between a guest and a parasite?' (59). Each website provides its own parameters for making this distinction, and for ensuring that its members stay on the 'guest' side of the divide. For example, when Ramon outlines his 'deal' to potential hosts, he is clear to his hosts and to his readers that he expects only a place to sleep and a meal, and preferably an Internet connection so that he can fulfil *his* part of the contract – which is to publish a story about his host on the *LetMeStayForADay.com* website. For a while, Ramon also incorporated a gift exchange into his project, asking each host to send a little something that he could give to the next host. Eight months into the trip, however, he had to forego the gift exchange when hosts started packing him off with objects that were too bulky or fragile, such as guitars and china teacups. Elsewhere on his website, Ramon's readers debate the other gestures that need to be factored into the give-take equation. One reader posts: 'Hey, I hope you were not only freeloading on others! What did you do in return?' Some of Ramon's critics wonder on the discussion forum whether Ramon says 'thank you' often enough to his hosts, or whether he is doing enough dishes or helping out enough around the house. Other readers, often past hosts, usually chime in that Ramon *has* said 'thank you' or *did* do the dishes, even if he did not write about it in his report. In these interchanges, the online community polices Ramon's 'deal' and ensures that the encounters with his hosts are reciprocated with the proper amount of gratitude and helpfulness, if not monetary compensation.

On the other websites, such as *CouchSurfing.com* and *GlobalFreeloaders.com*, the social pact of hospitality is itemized in similar terms. For example, in addition to stating what hosts are expected to offer (namely, a place to sleep for the night), *CouchSurfing.com* clarifies the guests' obligations. The following sample of tips on how to be a 'good' CouchSurfer highlights the fundamental basis of reciprocity, and hints at some of the anxieties this contract entails:

- As a good example of a good Surfer, you do as much as you can to give back to your hosts. This includes doing things like, for example, the dishes or stacking some wood. Maybe you have a special skill that you're willing to share.
- Bring a bottle of wine, or a six-pack of beer (go with the micro brew if at all possible, make it at least appear that you are making an effort; since you don't have to pay for a hotel room or hostel, cough up the $6–10). The wine is usually a safer bet than beer.
- Take a shower even if you took a shower the night before. One of the most important elements of being a good couch surfer is always appearing that you have somewhere to go… if people see you as a drifter with no direction, they will be a little worried about the chance of you trying to camp out on their couch longer than they would like you to.
- Nothing is better for a couch surfer than doing the dishes, a role 90 per cent of the population disdains. Either before you go to sleep, or when you wake up at another person's house first thing in the morning, do the dishes.
- Do not put your stuff in the bathroom or take up much space …The more care you take in respecting your host's space, the more your host will appreciate your company and be willing to host another surfer after you're gone.
- Always, always send a 'Thank-You' postcard from home!

These friendly tips serve to regulate the face-to-face encounters between members in the moment of offering and receiving hospitality. They point to the invisible boundaries between being a visitor or a drifter; between being a good guest or one who overstays their welcome. The specific material and embodied exchanges – hosts providing beds and meals; guests reciprocating with bottles of wine or by keeping themselves tidy and showered – are about the interpersonal relations of giving and taking hospitality. However, the websites also monitor how individuals give and take as members of the community. For example, the tip on respecting one's host so that they will be willing to host another surfer in the future underlines each traveller's responsibility to positively represent all other travellers in the community.

The very existence of hospitality exchange organizations relies on one basic condition: if you travel as a guest, you must also be willing to be a host. For example, part of Ramon's 'deal' is an open invitation to anyone who hosted him to stay at his house once he returns back home to Zwolle. Members of *The Hospitality Club*, *Servas*, *GlobalFreeloaders* and *CouchSurfing* are expected, and in some cases required, to be both hosts and guests as a condition of their membership in the community. As the *GlobalFreeloaders* site says: 'If there are too many people taking (visitors), and not enough giving (hosting), the system simply won't work.'

In order to ensure the proper functioning of the hospitality community as a whole, the websites necessarily blur the distinction between hosts and guests: hosts are just travellers who are not currently on the road, and travellers are hosts who happen to be travelling. As *The Hospitality Club* website states:

> The idea of The Hospitality Club is to give and to receive help…We don't ask you to do anything. We just offer you the opportunity to meet exactly the people that are going to help you on your next trip and exactly those people that will need your help when they visit your home country.

In this sense, reciprocity imposes a kind of solidarity across the community. All of the members have been, are currently, or soon will be travellers in need of a bed; likewise, all of the members are expected to be able to offer hospitality at some point. In this way, reciprocity becomes a measure of exclusivity; a way of binding those internal to the community while excluding those who are unable or unwilling to reciprocate properly.

In other ways, the websites naturalize reciprocity as an inevitable attribute of membership in the community. This is done in part by portraying hospitality as *its own* reward. Hosting others is not described as a chore, but rather as a fun way of being a tourist in your own city or living room. As the *CouchSurfing.com* website tells its members: 'By visiting someone you are allowing them to have a vacation even though they are at home.' Giving back by hosting is not seen as an obligation, but as a continuation of the pleasures of travel, as the following testimonials from hosts on *GlobalFreeloaders.com*, attest:

- Hosting is a great way of meeting people, great conversation, learning new and interesting things from around the globe. Puts excitement into a rather normal day. If you get the chance to host, don't miss it, you'll love it as we have.
- We were particularly looking for overseas travellers, so that our children could experience other cultures, languages & food.
- I love this site, you appreciate things so much more when you spend time with a traveller… So here I was, with a Canadian girl and a German girl in my

home, all at once. I have never had so much fun in my life, all these different cultures mixing at the same time.

- I was travelling without a backpack this time!

There are several different ways of thinking through these perceptions of the inherent rewards of hospitality. In one sense, the inherent rewards of hospitality appeal to a cosmopolitan fantasy that infuses all of these websites. The notion of getting close to the other – so close as to be invited into the stranger's home or to bring the stranger into your own living room – is central to cosmopolitan desires of consuming difference. As these comments suggest, the real thrill of hosting travellers is learning about, interacting with and mixing different cultures, even without having to pack your suitcase! In another sense, by highlighting these inherent rewards, the websites also mitigate the risk that taking will exceed giving. As long as the 'right' kind of people are included in the community, hosts need not worry that their generosity will be exploited by a freeloading drifter. Instead, they can expect the kind of 'reward of cosmopolitanism' expressed in these testimonials. But how do website ensure that they include the 'right' kind of people and put the 'right' kinds of difference into circulation within the hospitality community?

One of the key conditions of membership in these communities is the reciprocal exchange of trust. As the *CouchSurfing.com* website advises its members: 'Ask trust in return for your trust.' Trust is notoriously difficult to establish in online settings, and regulating trust in contexts of hospitality is especially complex. Thus, a second way that these websites ensure the proper relations of hospitality is by coordinating systems of safety and security within this context of risk.

Reputation

Benveniste highlights another indeterminate definition in which *hostis* can be understood as either 'guest' or 'enemy'. He says that 'to explain the connexion between "guest" and "enemy" it is usually supposed that both derived their meaning from "stranger", a sense which is still attested in Latin. The notion "favourable stranger" developed to "guest"; that of "hostile stranger" to "enemy"' (Benveniste, 1973: 75). Hospitality is always a risky affair, fraught with the anxiety that the guest may become a parasite, or worse, the enemy.

As with the ironic deployment of the term 'freeloader' on his website, Ramon also raises the fear that the guest may be an enemy with an image that recurs in several places on his website, and that graces the cover of the book he published about his journey. The image is a photograph of Ramon hitchhiking at the side of the road with a hand-painted sign that says 'I Don't Kill'. Ramon takes the risks associated with the stranger to their logical extreme, tapping the anxieties associated with hospitality in general, and especially hitchhiking. Ramon's sign raises fundamental questions: How can we know whether the stranger will be 'favourable' or 'hostile'? Whom can we trust and what constitutes trustworthy information about the strangers we might meet on the road?

Trust is as crucial to the effective functioning of hospitality communities as reciprocity, and the websites seek to circulate trust as a way of ensuring safety in the physical encounters between members. This is done through a variety of security systems that combine online information with physical verification. Among these are systems for verifying members' identities, face-to-face interviews between existing members and potential members, and the availability of identity profiles that hosts can match against the traveller who shows up at their door.

The most extensive security system that these websites operate is the 'reputation system'. Reputation systems, which are common in commercial websites like eBay or Amazon marketplace, are like short-cuts for establishing trust between strangers in online settings. As members have dealings with each other – whether online or in an embodied hospitality encounter – they can provide comments about each other to build up individual members' 'reputations'. For example, after having a traveller come visit, a host might add to that traveller's profile positive or negative comments about their behaviour as a guest. Similarly, the traveller can post comments about the host. These comments constitute the member's reputation and can be made visible to all other members.[3]

Resnick et al. (2000) explain how online reputation systems work: 'A reputation system collects, distributes and aggregates feedback about participants' past behaviour.… [T]hese systems help people decide whom to trust, encourage trustworthy behaviour, and deter participation by those who are unskilled or dishonest' (45–46). In their study of eBay and similar peer-to-peer retail sites, Resnick et al. attribute the counterintuitively high rate of success on eBay to the use of reputation systems and the ability of these systems to foster trust in the absence of other factors. Without face-to-face interactions, known histories, the prospect of future interactions, and social consequences for good or bad behaviour, reputation systems lend 'the shadow of the future to each transaction by creating an expectation that other people will look back on it' (Resnick et al., 2000: 46). Reputation systems establish histories for members and make these accounts of past actions visible to all other members. Future interactions can then be established based on these reports of past behaviour. This form of interpersonal surveillance within the online community disciplines members' behaviour both online and offline, ensuring that individuals act properly as hosts or guests and punishing them when they do not.

The discourse surrounding the reputation systems on the hospitality websites intimates that they are in place more to ensure the physical safety of their members

3 As Derrida (2000) reminds us, all gestures of hospitality must necessarily entail some risk. It is striking that these websites negotiate risk by invoking the notion of 'security' in terms of trust and reputation precisely at a time when debates around national security are dominated by sentiments of suspicion and disbelief (see Gibson Chapter 9 in this volume). Yet, it is important to note that discourses of domestic hospitality (as evidenced in these websites) are intricately related to the discourses and policies of national security that proliferated following the terrorist attacks of 11 September 2001. For one thing, the terrorist attacks in New York had repercussions for Ramon's project, as several of the New York residents who had signed up online to host him were, for various reasons, unable to fulfil their offers after the terrorist attacks. More generally, offers to host the traveller in one's home are contingent on the nation first welcoming the traveller across its borders, a welcome that is increasingly regulated.

than to complain that a guest left the bathroom a mess or that a host only offered white wine at dinner. One of the most common themes on each website's 'Frequently Asked Questions' page is safety: is it safe to go stay with a stranger or invite one to crash on your couch?

The *CouchSurfing.com* website operates interrelated forms of reputation systems that act as gatekeepers to the core, and presumably 'safe', group of members. First, the site uses a system of 'References' and 'Friend Links'. Anyone can leave a reference for any other CouchSurfer, building up each member's reputation. Friend Links allow travellers to associate themselves with other reputable travellers. As the website explains:

> Every user is linked to the other users he/she knows in the system through a network of References and Friend Links. These features help other users determine how trustworthy you are, based on the quantity and 'quality' of the people you know and also if you've been vouched for.

Second, *CouchSurfing.com* employs a system of 'vouching' that safeguards access to the 'core' network of friends:

> **Here's how it works!**
> Vouching on CouchSurfing.com is a security system. It signifies an elevated level of trust for those members who have become vouched for. To become vouched for, someone who is already vouched for must vouch for you…
>
> **Core Network**
> You can only be vouched for if you are connected to the original 'core' network of couch surfing members. It is that 'core' network of friends (connected through friend links) that is the vouched for network. How do you become vouched for if you are not part of that network? Easy. Go and surf with someone who is vouched for. After they meet you and get to know you, they can vouch for you.

Finally, for a fee of $25, members can also get 'verified'. The verification process is an identity check that substantiates the name and address that the individual provides. The reputation systems and other security systems that these hospitality websites put into place have two related effects. First, they bind the community even closer together as a 'safe' community in which the community is responsible for keeping itself safe. Resnick et al. note that 'as a solution to the ubiquitous problem of trust in new short-term relationships on the Internet, reputation systems have immediate appeal; the participants themselves create a safe community' (48). For example, members of *The Hospitality Club* are encouraged to report on each other for the safety of all:

> Members leave comments about each other on their 'profiles'. There you will be able to read the comments of other members about this member and see his/her past guests and hosts, as well as people that trust him/her. Please do write comments about other members yourself – this feature adds security for all of us.

Reputation systems are described through this rhetoric of safety of the community, but their second effect is to produce the community as an exclusive site of belonging. These security systems reproduce hospitality as an inherently risky affair; but rather than embracing that riskiness as Derrida advocates in his writings on hospitality, these websites work to mitigate and contain it, which results in a closed and exclusive community. These reputation systems work to keep the hospitality community openly closed, leaving us with a paradox that I want to discuss in the next section.

The Paradox of Global Community

By using the Internet to administer systems of reciprocity and reputation, these online hospitality communities help to alleviate the inherent risks of hospitality and ensure the smooth logistical and interpersonal operation of the hospitality exchanges. But the Internet is integral to these communities beyond logistics, surveillance or security. In fact, these online/offline communities claim that using the Internet for hospitality in this way gets back to what the Internet was *meant* to be used for. These hospitality sites hark back to the early principles of non-commercial, grassroots, democratic peer-to-peer communication and community, thus fulfilling the original utopian promise of the Internet to unite strangers across geographical and cultural divides and to form a truly global community. Ramon explicitly denounces commercial uses of the Internet on his website and Casey, the founder of *CouchSurfing.com*, was inspired to create his non-commercial site after becoming disillusioned by his lucrative career in software development during the dot-com boom. The not-for-profit status of the sites is inextricably tied into this formation of community, as *The Hospitality Club* website indicates:

> The Hospitality Club is a non-commercial project. We founded it because we truly believe in the idea that bringing people together and fostering international friendships will increase intercultural understanding and strengthen peace. We do not want to make a profit with this site.

By rejecting profit models and commercial exchange, these websites reassert the 'true' intentions of the Internet: to create a global village of strangers meeting strangers, sharing cultures and opening doors, hearts and minds. This utopian rhetoric can be found not only in the content posted by the website administrators, but also in posts by members, such as the following comment left on the *CouchSurfing.com* website:

> This is not just good use of the Internet. This is not just good use of a couch. Essentially what we have here is humanity working at its finest. People helping people, with no strings attached. I've only hosted a handfull [sic] of times, but each time changed my life for the better… Couch Surfing makes me belive [sic] that humans are essentially pure, good, and curious to know more about each other.

This comment and others like it reiterate some of the rhetoric surrounding the Internet and the virtual communities that were forming on bulletin boards and multi-user domains in the early 1990s, such as the CommuniTree Group analysed by Allucquère Rosanne Stone (1996) or The Well, an online community described in Howard Rheingold's book *The Virtual Community* (1994). Utopian thinkers at the time suggested that not only could virtual communities replace the sense of belonging that was missing in modern social life, but they could form better, more democratic and all-inclusive communities.

In her analysis of online feminist communities, Irena Aristarkhova (1999; 2000) suggests that Internet communities were never going to fulfil their utopian promise of an all-inclusive global community, partly because community is by definition exclusive. She notes that:

> Historically net communities have been based on protection and surveillance of their limits through the use of policing techniques that include censorship and exile of deviant users...The net communities were not even in their heydays spaces for free access, play and negotiation. They were from the very beginning 'governed' (another often-forgotten etymological meaning of the term 'cyber' – *kubernare* as in 'govern') spaces with clear notions of *propriety* (the 'dos' and the 'don'ts') and *property* (rightful ownership) (1999: 17).

Aristarkhova argues that 'closure is an essential and even constitutive gesture of net communities AND as such would always remain a stumbling block for those who conceive utopian visions of borderless and all-inclusive virtual communities' (1999, 18, original emphasis). She identifies an algorithmic logic in which net communities are formed by cancelling out individual difference:

> In all gestures to unity (unify) there is implied a *principle of homogeneity*, whereby things are submitted to an equation that cancels out their individual differences so that a larger unity based on some chosen feature(s) of 'sameness' could be *forged* (1999: 18).

Aristarkhova goes on to explain that community based on consensus or unity only pays attention to difference in order to eliminate or assimilate it: 'As such, the erection of a community is inherently allied to the construction of a defence mechanism which is vigilant to and exclusive of SOME other as foreigner and outsider' (1999: 19, original emphasis). In response to such exclusion, Aristarkhova turns to Derrida's notion of hospitality to challenge the logics of fusion and solidarity within the formation of community. She explains that for Derrida, communities are essentially *inhospitable structures*. However, 'lodging hospitality as a deconstructive *graft* within this structure of the community promises to keep it open to/for others. Hospitality allows communities "to make their very limits their openings" and thus ensures that there is always a possibility of hosting the other' (Aristarkhova, 1999: 19). For Derrida, hospitality does not exclude difference, but rather 'perpetually responds to the ethical demands of the heterogeneous, that is, of "others"' (Aristarkhova, 1999: 1). In other words, Derrida's concept of hospitality intervenes exactly at the point where communities reject, eliminate or exclude difference by offering an alternative for thinking about openings precisely where communities would otherwise erect boundaries.

So where does such an intervention leave online communities dedicated precisely to the arrangement of hospitality? Do these online communities eliminate or assimilate difference, as in Aristarkhova's description of fused communities?

Or does their ethos of hospitality and appeal to ideals of global community allow them to remain open to difference in ways that other communities are unable to do? Derrida's formulation of absolute hospitality relies on an unconditional opening up to the unknown stranger. This unconditional hospitality is offered without knowing the stranger's name or identity, without expectations of repayment, and regardless of the risks the stranger might pose (see Derrida, 1999a, 2000). In fact, Derrida argues, hospitality offered to a known stranger or in the expectation of repayment is not hospitality. Derrida critiques conditional models of hospitality based on reciprocity, such as Kant's ([1957 [1795]) notion of cosmopolitan hospitality. In reference to Kant, Derrida writes: 'to be what it "must" be, hospitality must not pay a debt, or be governed by a duty: it is gracious and "must" not open itself to the guest [invited or visitor], either "conforming to duty" or even, to use the Kantian distinction again, "out of duty"' (Derrida, 2000: 83). For Derrida, hospitality based on reciprocal exchange cannot be hospitality. Similarly, hospitality offered in the absence of risk can also not be hospitality. Derrida writes: 'Pure, unconditional or infinite hospitality cannot and must not be anything else but an acceptance of risk. If I am sure that the newcomer that I welcome is perfectly harmless, innocent, that (s)he will be beneficial to me… it is not hospitality. When I open my door, I must be ready to take the greatest of risks' (Derrida, 1999b: 137 quoted in Rosello, 2001: 11–12).

Contrary to these criteria, hospitality websites are instead based precisely on an economy of reciprocal exchange, both in the moment of the hospitality encounter and across the community as a whole. This reciprocity serves as a point of solidarity that bounds rather than opens up the community. Furthermore, reputation systems and systems of surveillance operate through the websites precisely to mitigate the risks posed by welcoming strangers into one's home. The hospitality on offer in these website communities is, if anything, conditional and contained. The reciprocity between hosts and guests in the moment of hospitality, and more generally within the community as hosts become guests become hosts, forges a global community in which members are open to each other, but enclosed within this economic chain of reciprocity and obligation. The reputation systems serve to further police the boundaries of this community, partly under the guise of ensuring the safety of members from each other by excluding or alienating 'questionable' subjects, but also as a way of constructing bonds amongst kindred spirits; of creating an enclosed cosmopolitan community – paradoxically, a closed community of open-minded and like-minded people.

However, if these hospitality communities are not entirely open to difference, neither do they necessarily eliminate difference. The fundamental themes of the hospitality websites – world travel and hospitality – reflect a cosmopolitan desire for and openness to difference. In this sense, the communities do not reject difference, but rather 'filter' it in order to allow the community to internalize the 'right' kind of difference while excluding the 'wrong' kind of difference.

The Paradox of Cosmopolitan Hospitality

Ramon's website and other hospitality exchange websites revolve around cosmopolitan fantasies of proximity to the other, global community and the 'wealth of difference'. Sentiments similar to the ones expressed by these *CouchSurfing* members infuse all of the websites:

> Couchsurfing is another amazing thing that happen in my life. This makes me realize that people still believe into a unified world and that differences are our wealth and we are winner to learn from each other.
>
> It opens the door to unimagined adventures, and special bonds with people all over the world. Congratulations to the founds and all the people who contribute to the success of this global community.

How does someone become a member in this global community? What kinds of qualities or behaviours ensure that you will be welcomed into the community and get a 'good reputation' once you are there? Despite claims that 'everybody is welcome' (*The Hospitality Club*), it quickly becomes clear that this is not entirely the case. These website implement several conditions of membership, a few of which I have already discussed, such as an ability to reciprocate both in the hospitality encounter and to the community at large, a verifiable identity and a 'clean' profile, or prior inclusion in a face-to-face social network. Another important condition that members must meet, however, has more to do with one's attitude. For example, the *Servas International* website asks:

> Are you:
> Friendly?
> Curious?
> Open-minded?
> Would you like to visit foreign countries and take part in everyday life?
> Would you like people from other countries to join in your daily life for a short time?
> Do you try to overcome your prejudices to communicate with others?
> Do you believe that peace is possible if everyone truly wants it?
> Then SERVAS is for you.

As this checklist suggests, these communities require their members to have a cosmopolitan disposition to the world and, specifically, to difference. Members are expected to be curious and open-minded. They are expected not just to tolerate difference, but to celebrate it

In some websites, this cosmopolitanism manifests as what Urry (1995) refers to as 'aesthetic cosmopolitanism', a model that sees the cosmopolitan as a highly mobile, curious, open and reflexive subject who delights in and desires to consume difference. In other instances, this cosmopolitan sensibility takes on a broader political significance related to utopian ideals of global community and world peace, as we see in the quotes above and in the following statement on *The Hospitality Club* website:

> The club is supported by volunteers who believe in one idea: by bringing travelers in touch with people in the place they visit, and by giving "locals" a chance to meet people from other cultures we can increase intercultural understanding and strengthen the peace on our planet.

Whether aesthetic or political, the cosmopolitanism described in these websites is fashioned through proximity to the other. It is through encounters with the stranger that differences are produced, shared, consumed and, in the utopian ideal, transcended to forge a global community. But this proximity is risky. To be close to the stranger is to always be vulnerable to the stranger, and to the possibility that the stranger will be too different or not different in the 'right' way. And yet, being close to the stranger is necessary to define the community's identity first as a community, that is an exclusive and bounded space of membership, and secondly as a global community, that is a community whose identity is intimately bound up in fantasies of difference.

However, as Sara Ahmed reminds us: 'While identity itself may operate through the designation of others as strangers, rendering strangers internal rather than external to identity, to conclude simply that we are all strangers to ourselves is to avoid dealing with the political processes whereby some others are designated as *stranger than other others*' (2000: 6). In the websites, some strangers are internal to the community in the sense that they provide the members with the wealth of diversity that makes strange encounters rewarding and defines the group's global and open identity. These strangers are seen as friends. A frequently quoted motto on the *Global Freeloaders.com* website is: 'A stranger is just a friend you haven't met yet'. *The Hospitality Club* refers to its database as a 'World Wide Web of friendly people' and the *Hospitality Exchange* website similarly positions its community as a group of friends among whom:

> common themes immediately jump out – love of traveling, of course, but also… our members' passion for food, conversation, music, outdoor activities of every kind, and sharing local attractions with other travelers… Look at the sample listings to see the kind of kindred spirits you'll discover in the Hospitality Exchange community of travelers and friends.

Here, difference between people is managed through broader commonalities that unite the community in solidarity: 'kindred spirits', like-minded people and an emphasis on common interests in travelling, learning about other cultures and eating foreign foods. Stranger-ness fulfils the community's need for differences that are consumable *and* communal.

The 'favourable stranger', then, is merely a friend you haven't met yet. But what of the 'hostile stranger'? How do 'other others' contribute to the community's identity precisely through their exclusion from it? Other strangers, whose difference threatens rather than serves the cosmopolitan fantasy, are invisibly present on these websites, often in the fine print. For example, near the bottom of its list of 'Rules', *The Hospitality Club* includes the following items:

> Do not spam!
> Do not send messages to another member that are not related to the aim of The Hospitality Club. Those are especially but not exclusively requests for help finding work, apartments and visa invitations.

Clearly, people who do not already have the financial means to travel, a place to host other travellers or the political right to mobility are not welcome to participate in the club. This admonition hints subtly at the fact that asylum seekers in need of visas or strangers who intend to visit and *stay* (by getting jobs and apartments) are excluded from the kind of hospitality on offer here. Guests who might become parasites or enemies represent the 'wrong' kind of difference; a difference that is not easily consumed over a glass of wine or a late night conversation in someone's living room.

One of the most fundamental, and yet completely tacit, assumptions these websites make is that their members are already full-fledged global citizens whose middle-class status and 'First World' passports ensure that nations will happily extend hospitality to these travellers. It is upon this basis, then, that the participants can claim to be forging an open global community, albeit one that is already closed before it even gets started.

Conclusion

I want to conclude with some comments about another cosmopolitan fantasy that circulates on these websites, the notion of 'being at home in the world':

> The only requirement is that you open your heart and your home to hosting others, and the whole world becomes your extended home! What a way to promote the values of world peace and cultural understanding, something that is needed now more than ever in the world we live in!
>
> (*GlobalFreeloaders.com*)

> Bring the World home! (*CouchSurfing.com*)

Drawing on Dikeç's (2002) analysis of spaces of hospitality, and his references to Honig's (1999) arguments on hospitality and home, I want to ask whether the cosmopolitan fantasy of being at home in the world might be recuperated as an opening for hospitality exchange communities. To be sure, the rhetoric on the websites of bringing the world home and being at home in the world is underpinned by problematic associations of hospitality with rights of property and of 'feeling at home' with a kind of white, wealthy, Western privilege that allows travellers to feel comfortable wherever they are. But what if being at home in the world was not about comfort and not about rights of property; and what if we thought of the space of home not as static and secure, but rather as mobile and, as Dikeç argues, disturbed?

In his critique of discourses of national hospitality toward immigrants, Dikeç asks:

> Isn't it timely to engage and challenge, as Honig suggests, 'the seduction of home', the seduction of the 'construction of "homes" as spaces of safety, spaces safe from the disturbance of the stranger'? Isn't it timely to consider the usurpation of speaking the language of hospitality in order to construct safe homes? Isn't it timely, in short, to reconsider the notion of hospitality, to reconsider what it means to be host and guest, to be disturbed, as Levinas once hinted at, by 'being at home with oneself'? What if being disturbed by 'being at home with oneself' turns into being disturbed by the stranger? (2002: 242).

Dikeç advocates, instead, a notion of home as already disturbed by the stranger. This calls for a radical rethinking that allows us to decouple home from stasis, and mobility from disruption, and instead conceive of home as already mobile, already disturbed by strangers (see Ahmed, 2000; Fortier, 2003). As such, 'being at home in the world' would refer to an openness that resists the urge for safety and comfort. It would embrace, instead, the risks as well as the positive possibilities inherent in Derrida's (2000) notion of absolute hospitality.

Perhaps another way of recuperating the cosmopolitan fantasy of being at home in the world is to rethink Kant's (1795) notion of cosmopolitan hospitality in which hospitality is framed as the visitor's right and the host's obligation. For Kant, the laws of cosmopolitan hospitality assure the right of resort to the stranger; but what if the law of cosmopolitan hospitality assured, instead, the right to offer hospitality, in addition to the right to receive it? If we refigure 'home' against conditions of property, ownership, safety or enclosure, and instead in terms of openness, mobility, strangeness and risk, then perhaps we can image a world in which *everyone* can feel at home and at home with oneself precisely by openly welcoming the other.

References

Ahmed, S. (2000), *Strange Encounters* (London: Routledge Books).

Ahmed, S., Castañeda, C., Fortier, A.-M. and Sheller, M., eds (2003), *Uprootings/Regroundings: Questions of Home and Migration* (Oxford: Berg Publishers).

Aristarkhova, I. (1999), 'Hosting the Other: Cyberfeminist Strategies for Net-communities', in Sollfrank and Volkart (eds).

—— (2000), 'Otherness in Net-Communities: Practising Difference in Post-Soviet Virtual Context', Paper presented at Situating Technologies Symposium, De Balie, Amsterdam, 1 April 2000.

Benhabib, S., ed. (1999), *Democracy and Difference: Contesting the Boundaries of the Political* (Princeton, NJ: Princeton University Press Books).

Benveniste, E. (1973), *Indo-European Language and Society* (London: Faber and Faber Limited).

CouchSurfing.com [website], <http://www.couchsurfing.com>, (accessed: 14.08.06).

Derrida, J. (1999a), *Adieu to Emanuel Levinas*. trans. Brault, P.-A. and Naas, M. (Stanford, CA: Stanford University Press Books).

—— (1999b), 'Dèbat: Une hospitalitè sans condition' in Seffahi (eds).

—— (2000), *Of Hospitality, Anne Dufourmantelle Invites Jacques Derrida to Respond*. trans. Bowlby, R. (Stanford, CA: Stanford University Press Books).

Dikeç, M. (2002), 'Pera Peras Poros: Longings for Spaces of Hospitality', *Theory, Culture & Society*, **19(**1–2), 227–247.

Fortier, A.-M. (2003), 'Making Home: Queer Migrations and Motions of Attachment' in Ahmed et al. (eds).

Gibson, S. (2003), 'Accommodating Strangers: British Hospitality and the Asylum Hotel Debate', *Journal for Cultural Research*, **7**(4), 367–386.

GlobalFreeloaders.com [website], <http://www.globalfreeloaders.com>, (accessed: 21.08.06).

Honig, B. (1999), 'Difference, Dilemmas, and the Politics of Home' in Benhabib (eds).

Hospitality Exchange [website], <http://www.hospex.net>, (accessed 18.08.06).

Kant, I. (1957 [1795]), *Perpetual Peace* (Indianapolis, IN.: Bobbs-Merrill Books).

LetMeStayForADay.com [website], <http://www.letmestayforaday.com>, (accessed 09.07.05).

Resnick, P. et al. (2000), 'Reputation Systems', *Communications of the ACM*, **43**(12), 45–48.

Rheingold, H. (1994), *The Virtual Community: Finding Connection in a Computerized World* (London: Secker & Warburg Books).

Rosello, M. (2001), *Postcolonial Hospitality* (Stanford, CA: Stanford University Press Books).

Seffahi, M., ed. (1999), Manifeste pour I'hospitalitè, aux Minguettes: Autour de Jacques Derrida (Gringy: Paroles d'aube).

SERVAS International [website], <http://www.servas.org/ >, (accessed: 15.08.06).

Sollfrank, C. and Volkart, Y., eds (1999), *Next Cyberfeminist International Reader* (Hamburg: Old Boys Network) [website], <http://www.obn.org/obn_pro/downloads/reader2>, (accessed: 12.08.05).

Stone, A.R. (1996), *The War of Desire and Technology at the Close of the Mechanical Age* (Cambridge, MA: MIT Press Books).

The Hospitality Club, [website]. <http://www.hospitalityclub.org/>; (accessed: 18.08.06).

Urry, J. (1995), *Consuming Places* (London: Routledge Books).

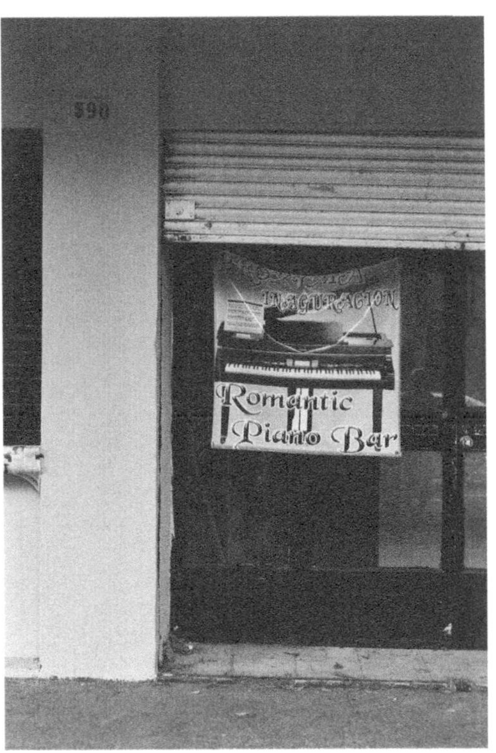

Romantic Piano Bar (Guadalajara, Mexico, 2006)

Chapter 5

Sensing and Performing Hospitalities and Socialities of Tourist Places: Eating and Drinking Out in Harrogate and Whitehaven[1]

Viv Cuthill

This chapter explores the intersections of mobilities, commercial hospitalities and socialities performed in tourist places, and suggests that notions of hospitality and place are sensed through individual and group entanglements in the complex flows of people, food- and drink-scapes, capital, images, objects and fashions found in eating and drinking venues.

Cafés, restaurants and bars are commercial hospitality spaces where tourist places are tasted by visitors and residents. Snapshots of place are gazed upon, sensed and collected at lunch in a restaurant or over a drink in a bar. Simultaneously, the hospitalities created in venues communicate messages of style and taste to different communities of consumers. Visitors and residents participate in embodied consumption performances in food and drink spaces, so that cultural codes and social meanings develop and are performed in venues.

I propose that service cultures in eating and drinking venues are created, and momentarily concretized and sensed, at the intersection of the imagination and performance of commercial hospitalities and visitor and resident socialities in specific material and physical environments. Service cultures produce senses of place and reflect cultures of place at particular moments in time. I illustrate this by deconstructing service cultures in two commercial eating and drinking venues in two northern English tourist towns, Harrogate and Whitehaven.

Harrogate, a middle class spa town and conference centre, is a long-established tourist destination. Whitehaven, a working class port, former mining town and market town located immediately north of the Sellafield nuclear plant, is a recently developing tourist place. These peripheral tourist towns, less well-known places that are relatively geographically and culturally marginal, demonstrate how tourism

1 Thanks to Ray Cotgrave, Manager, Revolution Vodka Bar, Harrogate and Ricky Andalcio, Owner/Manager, Zest Harbourside, Whitehaven for participating in interviews. Thanks to Jennie Germann Molz and Sarah Gibson for their patience and for pointing me in the right direction.

development pulls places into global tourist circuits and accelerates the flows of fashions and tastes in eating and drinking into and out of place.

Change in eating and drinking venues raises questions about the hospitableness of place, and the place of hospitality. When commercial hospitality spaces are designed to attract visitor markets, to what extent are different resident groups welcomed into venues, and how welcome are visitors in established resident venues. It is through performances of service cultures that commercial hospitality spaces communicate messages of inclusion and exclusion for different social groups.

Eating and Drinking Place

A plethora of cities, towns and villages are being redefined as centres of cultural consumption and re-presented as tourist places. Carefully projected images of place create a desire and a demand for travel and attract visitors seeking new experiences. International consumer services and brands flow in and out of high streets as people flow in and out of new centres of consumption. Consequently, as everyday lives become inextricably wired to global networks of tourist mobilities (Sheller and Urry, 2004), it is impossible for places to be disconnected culturally from the wider world. It is interesting to explore, therefore, how global connections of tourist mobilities influence the lived experience of place.

The proliferation of urban lifestyles witnessed in recent years has fuelled the cultural competition between urban areas aiming to attract internationally mobile residents, visitors and investors. Zukin (1998) suggests that the meaning of urban lifestyles now refers to an aggressive pursuit of cultural capital that encourages various forms of cultural consumption in which highly visible consumption spaces such as restaurants and bars are key. Consequently cafés, bars and restaurants flow in and out of places signalling messages of style and taste. Bell (2006) has demonstrated how food and drink spaces add to the attractiveness of cities and symbolize regeneration and a new confidence in Manchester, England. He argues that contemporary cityscapes are maps of 'distinction'. The city is a map of culinary tastes, divided into taste zones, with taste-makers driving the system of distinction. Thus, places are judged, ranked and repositioned as desirable or undesirable based on perceptions of the figurative and literal tastes that can be sampled and expressed in place. Examining urban cultures created by performances of hospitalities and socialities in eating and drinking venues therefore enables an analysis of some of the consumption and image factors driving global place competition. It also highlights the mobilities and practices of the chasers and definers of cultural capital in place, who weave desirable places and covetable eating and drinking activities in place into webs of stories and narratives they produce when constructing their social identities.

Eating and drinking out are particularly interesting entertainment and cultural consumption activities because they purvey multiple social, cultural and symbolic meanings that extend beyond preferences in food and drink. As Finkelstein (1989: 41) has identified, tastes in foods and preferences in the style of dining are not independent of other features of the social epoch and she views the restaurant as a symbol of economy and society. Pillsbury (1990: 11) agrees and labels the restaurant a cultural barometer, arguing that in contemporary society the restaurant is a mirror of ourselves, our culture and our new geography. The restaurant is also a nexus of complex mobilities and, as Zukin observes (1995: 159), the development of the restaurant industry is inextricable from global processes of change, the restaurant being a focus of transnational economic and cultural flows and a place where cultural products are created and reproduced. I agree and argue that it is not only in the restaurant, but in the shifting combination and complex mobilities of restaurants, cafés, pubs and café-bars in place, that global and local cultures are synthesized and cultural styles fused and diffused. It is thus in these venues that cultural change can be identified and interpreted, as shifting tastes in eating and drinking out reflect and reproduce features of the epoch.

Experiences of eating and drinking out have been acknowledged in the existing literature on performing tourist places (see Urry, 2002 and Baerenholdt et al., 2004), but have not been credited with the central role I believe they hold in the socialities of place. Individuals experience heightened sensations in eating and drinking venues that present a more acute, embodied, interactive place experience than strolling in the street. The sensing of the quality of food, drink, service and environment is central to the sensing and judging of place. Visitors and residents participate in different and overlapping performances. Residents embody a sense of their connection with place, visitors embody their tourist status. Resident groups identify with each other and with particular eating out and drinking out experiences that create meaning in place (Bell and Valentine, 1997). Tourists engage with spaces of hospitality away from home that represent place or that represent their sense of selves.

Visiting particular restaurants, cafés and bars expresses personal taste, and aesthetically reflexive consumers are particularly adept at expressing senses of self through participation in consuming activities. Crouch (1999) suggests that people who participate in leisure and tourism participate in critical negotiations of the self. He believes that the contemporary importance of lifestyle suggests that practice is messy and that there is negotiation and tactic in semiotic consumption, so that commercially designed sites provide metaphors through which to adjust, hold onto or discover aspects of lifestyle. Restaurants, cafés and bars are, therefore, ideal sites to engage with contemporary debates on the performance of lifestyle and taste. Observing a range of practices in and around these venues allows for interpretations about how people view themselves, the social relationships they participate in and how they identify with the place they are in.

I argue that social relations are performative, sensuous and fluid. I examine the socialities performed in and between commercial hospitality spaces in the public spaces of tourist places, as visitors and residents perform identities and place their tastes and urban play-spaces become coded according to manners and practices. I engage with the perspective of Maffesoli (1996) in his treatise on the decline of individualism in mass society. Whereas academic analyses of taste in modernity equated preferences of taste with conspicuous consumption, displays of status and class struggles for distinction (Bourdieu, 1984), I examine Maffesoli's claim that in performing multiple socialities in place, groups place their tastes, but these tastes are not primarily class-based and instead express the affective solidarity of the group (see also Laurier and Philo (2005) on the performances of 'cappuccino communities'). I investigate how people identify with or feel alienated by the ambience created in an eating or drinking venue, and therefore engage with experiences and places that ally them to certain taste communities, thus indicating a close connection between commercial hospitality, spatial experience, lifestyle and taste.

Maffesoli (1996) suggests that contemporary social life is organized around the membership of small groups, temporary groupings characterized by members' shared feelings, lifestyles and tastes. Theirs is an 'empathetic sociality'. People demonstrate their membership of these emotional communities through co-present performances, and thus become part of the urban landscape. He argues that the spirit of the times is found in sociality rather than the social, and that social puissance, the inherent energy and vital force of people, is overtaking the power formerly exerted by institutions. He is particularly interested in the ambience or aura of places that expresses the spirit of the times and suggests (1996, 6) that the ambience of the current era is built on a fundamental paradox: the constant interplay between growing massification and the development of micro-groups, or tribes.

Urban groups, therefore, can be distinguished and identified by the shared lifestyles and tastes of their members. The group actively creates its own myths and forms itself around specific rituals, shared values and ideals, types of dress and other group specific styles of adornment. As the ethos of the tribe evolves, certain rules and rituals develop, so that these 'organic' socialities connect and bind groups together to create meaningful social lives. Maffesoli (1996: 13) suggests that the empathetic sociality of these groups is expressed as ambiences, feelings, and emotions, and that the collective emotions of a group become concrete in places such as local pubs or in a 'network of convivial cells' (1996, 42). Group power, or social puissance, derives from being-together, and from the networks of friendships that derive from proxemics, or physical co-presence in place. Having a few drinks in a bar or a meal in a restaurant binds people together into their chosen communities, and restaurants, bars and cafés are important places of lifestyle display. I identify rituals, situations and experiences that relate to the socialities of specific affinity groups in Harrogate and Whitehaven to ascertain how group membership is expressed through acts of cultural consumption in eating and drinking venues.

Maffesoli's perspective is particularly useful for conceptualizing co-present group performances of eating and drinking out in place. Proxemics and the undirected 'being-together' are concepts central to Maffesoli's argument and these concepts are central to the sensing and performing the hospitalities and socialities of place. I focus on eating and drinking venues as social spaces in which affinity group socialities, lifestyles and tastes are performed. However, socialities performed by customers in eating and drinking venues do not operate in isolation from the hospitalities performed by staff, nor from the design of the physical and material environment, therefore I now conceptualize how service cultures in eating and drinking venues are created, imagined and performed.

Sensing and Performing Service Cultures

I propose that service cultures in eating and drinking venues are formed at the intersections of venue hospitalities and customer socialities. Initially the analysis of this proposition would seem to fit neatly into a production and consumption framework. However, I argue that conducting a separate analysis of the production and consumption of service delimits the discussion and is insufficient in identifying the multiple intangible moments of cross-over between service production and consumption in any service setting. The metaphor of performance, however, bridges that gap and brings together the active, immediate roles of staff and customers in creating a service culture through multiple performances of hospitalities and socialities.

Crang (1997) uses the metaphor of performance to encapsulate notions of working under the tourist gaze. In theorizing the nature of tourist-related employment, Crang highlights the performance of the tourist product and proposes a framework to theorize the social and spatial relations of provision, importantly identifying connections between tourism producers and consumers. He moves beyond assertions of the dramaturgical character of tourism work, to examine the constructions and contestations of tourism sites and their performed personae. Following Crang, I conceptualize a service culture, to encapsulate the active, sensuous role and co-present interaction of customers and employees with each other and with the texts, visual cues, and sensations in the service environment. I suggest that service cultures are created and sensed in eating and drinking venues through the commingling of the following features: the visual appeal of the material environment, the complex and mixed personal sensations of that environment, and through performances in a multitude of possible scenarios between customers and staff. Together these factors create an ambience specific to each venue.

Firstly, material culture and specific commodities create a visual image or impression of an eating and drinking venue and service culture. The visual impression is communicated externally by the architecture and appearance of the building and choice of venue name. The presentation and content of written incentives outside, such as 'buy one get one free', or a menu on public display will affect decisions to enter, as will the appearance of the interior from the exterior: whether the venue appears warm and cosy, lavish and opulent, or dark and dingy. The internal décor creates a visual impression of a service culture, such as a modern chrome minimalist café with designer chairs and tables, or an Italian restaurant with Chianti bottles and painted scenes of Italy on the walls. These are the scenic aspects of the setting (Goffman, 1956: 33).

In supporting the performance of the service culture, cutlery, crockery, glassware and other material objects all work in unison to create a particular impression, as does the quality and presentation of food or drink. Consider the visual effect of a work of fusion food art piled up in the centre of a large, chunky white plate, or the carefully presented and arranged numerous and varied dishes that create a Chinese banquet. Goffman (1956: 33) refers to these items as 'sign equipment' on the front stage that create the setting for performances.

Importantly, also, the visual impression of staff and customer clothing and appearance operate in tandem to create a visual image of a venue's ambience and service culture. Employees embody the culturally meaningful product and, according to Crang (1997), become the bodily habitus expected by tourists. Simultaneously, I propose that customers have to embody an appropriate customer status by displaying the bodily habitus that corresponds with the style of the venue, and then engage in practices appropriate to the service setting.

Secondly, material factors affect sensuous perceptions of the service experience. Maffesoli (1996: 31) claims we have left an optical period and are now entering a 'tactile' period. In addition to visual aesthetics customers sense the ambience of a service culture through other human senses: tactile, auditory, olfactory and gustatory. The feel of thick carpets, sticky linoleum, hard chairs, or leather sofas communicate subtle messages to consumers, as do sounds that assault the ears such as the choice and volume of music or absence of it. Sensory aesthetics and ergonomic design features play a significant role in the customer's emotional response to the physical environment. Simultaneously, customers imaginatively, visually and corporeally engage with food and drink products. Smells create instant snapshots of service cultures, as in the feelings evoked by the smell of burned fat or, alternatively, fresh bread.

The ambience of the service culture is felt almost imperceptibly through cumulative multi-sensory experiences, therefore, when snapshots and sensual memories of the venue are formed, they are associated with the geographical place in which it is located. As Crouch and Desforges (2003: 8) suggest, encounters with places and people are productive of a lay geographical knowledge that derives from the multi-sensual character of encounters though which people 'feel' their way in relation to cultural contexts.

Finally, interpersonal staff and customer performances affect emotional and aesthetic perceptions of the service experience. Warde and Martens (2000: 187) note that the immediate sensual pleasures derived from the embodied activities of eating and drinking, and actively engaging with a physical environment, are enhanced in the company of others. Verbal and non-verbal interactions between customers and staff, customers and other customers, staff and managers actively shape and maintain service cultures, whether a highly scripted performance, or an individualized service encounter. As Goffman (1956) observed, people must expressively sustain a definition of the situation through the tacit rules of face-to-face interaction, in the way people are greeted and served, the way the service environment is presented and controlled by staff, and in the conversation and behaviour of other customers as co-participants in the performance. The hospitalities of the venue are sensed in the combined performances of staff and customers.

Service cultures are created, therefore, through material, sensuous, performative and interpersonal factors that together produce distinctive service experiences. Identifying and deconstructing the different elements of an eating and drinking service culture illustrates how service cultures are composed at the intersections of complex networks of flows. Mobilities of food and drink objects, material culture, the materiality of buildings and performances of workers and customers, residents and visitors from different places combine to create service cultures.

Place is momentarily sensed in the combination of mobilities of objects, fashions and tastes, interacting with performances of hospitalities and socialities that create service cultures. Therefore, as mobilities of fashions and tastes in eating and drinking out enter place, and different people participate in service performances, they alter service cultures created at the intersections of venue hospitalities and socialities and modify senses of place.

Hospitalities and socialities in venues involve affective affiliations that play out through complex negotiations of inclusion and exclusion to communicate a sense of welcome and (in)hospitableness. When performances of particular socialities intersect with performances of specific hospitalities in eating and drinking venues, the service cultures created express the lifestyle choices of the people who frequent them.

To illustrate this point, I examine a particular category of contemporary fashionable eating and drinking experience: the stylized venue. I suggest that eating and drinking venues can be categorized according the genre of service they purvey. Whereas Warde and Martens (2000: 226) identified that in the 1990s familiarity with ethnic cuisine was a mark of refinement, I argue that in the 2000s eating and drinking out in carefully designed, 'stylized', often experiential (see Pine and Gilmore, 1999) eating and drinking venues is a contemporary marker of taste and cultural style. The arrival and mix of 'stylized' eating and drinking venues in place, symbolizes the contemporary desirability of place and provides the taste experiences desired by people on the move and taste communities in place.

I use the term 'stylized service genre' to refer to the service cultures of eating and drinking venues that are viewed as fashionable, ephemeral and in tune with the zeitgeist. The cutting edge design of venues marks them out as being in tune with contemporary cultural style. In Britain stylized venues include European style café-bars, modern cocktail bars, vodka bars, and gastro-pubs; restaurants with a cutting edge theme, often linked to celebrity chefs or celebrity owners with names like 'Fifteen' or 'Lime' or 'Mash'; chic restaurants and bars within city centre boutique hotels, bars and restaurants such as @ Home and Duvet, cafés in designer department stores, such as Harvey Nichols in Leeds, and restaurants in cutting edge visitor attractions such as Le Mont in Urbis in Manchester.

Venue design is self-consciously stylish. The décor, furniture and music will all be in vogue. The service style is equally self-conscious and is required to fit exactly with the theme of the outlet with staff performances, clothing, and appearance carefully prescribed. Stylized service can often be a de-humanized, impersonal style of service, with the server retaining control of the service environment and delivering a carefully prescribed performance. In current trends this relates to a European style of service where young waiting staff dressed in modern minimal black clothing, with fashionable hair styles provide a slick service with a minimum of conversation and few smiles, except with those customers who are known and accepted. Their role is not to be servile but to perform the image, ambience, and style of the venue. Hairstyles, physique, age, and demeanour all communicate a certain cultural capital that relates directly to the image of the venue, demonstrating the central role of aesthetic labour (Warhurst et al., 2000) to the stylized genre. Employees embody a 'sociality' demanded by both the organization and customers. Despite the seriousness of the style endeavour, however, the emphasis of the experience in many stylized venues is on play and fun.

Stylized service communicates a 'style' message to the consumer, and can be experienced in different ways. For the knowing consumer that falls within the target market range of particular outlets, the service experience can be fun, cutting edge, providing public access to the fashionable crowd. For the less knowledgeable customer, or one who stumbles into a stylized venue without realizing the strict, yet unstated, appearance and behaviour codes of a particular style environment, the feeling can be one of self-consciousness, discomfort or exclusion. As these outlets are often designed with a particular age group in mind, the feeling can be one of being too young or too old, certainly not one of the crowd. Even though nothing obvious indicates this fact, the server and other customers play a role in communicating to individuals verbally or non-verbally that they are 'out of place'.

Richardson (2000: 142), for example, is transported by 'a sky-blue painted proto-industrial lift' to 'Air', Oliver Peyton's 'gastro-paradise' in Manchester. This carefully designed environment is described by Richardson as a simulacrum of twenty-first-century glamorous living:

> High ceilings, white walls flooded with light from the big factory windows; curvilinear, deep blue, organic, somewhat seventies inspired furniture, with separate sofa units for extra-cosy groupings; flying saucer lamps. It was like a scene from *Sleeper*. Blue lacquer tables, and an eau de nil acrylic floor that squeaked a little as the waitresses made their way towards your table.

He orders two starters, and then wonders if such a request was too 'eighties' and that the waiting staff would be giggling about this behind their hands. The stylized service culture causes him to question his request making him feel out-of-date and, perhaps, a figure of fun.

Trends in stylized service genre venues usually begin in metropolitan areas and then move out to the regions. The stylized environments and service styles are not allied to the geographical places in which they are located; they produce mobile service cultures that operate independently of place. However, these venues synthesize global and local service cultures and are often instrumental in the diffusion and fusion of new cultural styles in geographical places (Zukin, 1995). Visiting stylized venues is clearly associated with placing taste, identifying with the style crowd and accruing cultural capital.

Service cultures performed within specific service genres, therefore, mark hospitality spaces out for particular types of people. Customers sense the hospitableness of a commercial eating and drinking venue by assessing the different factors that cumulatively create a service culture. Venue hospitalities are welcoming to those who fit in with the service culture, and are sensed as unwelcoming by those who choose not enter. Eating and drinking venues are not necessarily designed to feel hospitable and welcoming to all.

Whereas Telfer (1996) suggests that hospitableness is the giving of food, drink and accommodation to the stranger, and Derrida (2002: 358) contends that it is difficult to dissociate a culture of hospitality from a culture of laughter or a culture of smile, in commercial hospitality venues there are other factors linked to fashion, taste and identity that complicate these statements. I argue for a nuanced examination of enactments of hospitableness in commercial eating and drinking spaces through the prism of the service culture to explore how hospitable stylized hospitality spaces are for different resident and visitor groups.

Sensing Harrogate and Whitehaven

I now examine socialities and hospitalities performed in two northern English towns, Harrogate and Whitehaven. My analysis is based on group performances and participation in socialities and hospitalities in a stylized venue in each place: Revolution Vodka Bar and Zest Harbourside. I deconstruct the hospitalities, socialities and materialities that intersect to create the service cultures in each venue, and make connections between physical spaces and symbols, the built environment, sociability and urban lifestyles (Zukin, 1998) to develop an understanding of the intersections of urban performances of group socialities and eating and drinking hospitalities, and to identify customer inclusions and exclusions in each venue. I illustrate how performances of stylized service cultures create different senses of hospitableness and place in each town.

Harrogate is a tourist-historic town and a highly developed space of consumption. Whitehaven is a post-industrial town that has lost traditional industries and is turning to the development of tourism and service sector jobs to regenerate and support a depressed local economy. Differences are instantly evident in the built environment that reflects the different economic, social and cultural development histories of each town, and in contemporary consumer services that reflect each town's current economic status.

The trajectory of tourism development in each place has been completely different. Harrogate was an early English tourist destination, its development as a town evolving from its discovery as a spa in 1571. The fashion for visiting spas for health and leisure attracted large numbers of visitors to Harrogate in the nineteenth and early twentieth centuries. However, a decline in the demand for spa treatments after the Second World War, led to the redevelopment of Harrogate as an international conference centre. The local economy is and always has been a service economy, therefore Harrogate has always responded to external demands and fashions to continue to attract visitors to support its extensive tourism superstructure.

In tourism development terms Whitehaven is at a different end of the spectrum from Harrogate. Whitehaven is located on the North-West coast of England and is peripheral even to the North, both geographically and culturally. It is a Georgian town founded on the wealth generated by mining. By the mid-eighteenth century it was also renowned as the third busiest port in England. The twentieth century saw the loss of traditional mining, shipping and chemical industries and with the proposed decommissioning of the nearby Sellafield nuclear plant, tourism development is being used to create employment and attract investment. In the mid-1990s, economic development in the town received a boost when the Whitehaven Development Company attracted funding to develop the town as a tourist destination.

Therefore, Whitehaven is transforming from a de-industrialized market town for a rural community to a tourist place attracting visitors from around the globe.

Harrogate has a reputation as a place with traditional quality hotels, shops, restaurants and cafés and a service culture that reflects its spa town and conference town status, and the demands of a middle class, affluent population. The working class town of Whitehaven does not have a similar range of quality accommodation, shopping facilities and eating and drinking opportunities, hence Whitehaven does not have the image of being a place to visit or stay, or a place in which to eat or drink out. Dominant eating and drinking venues in Whitehaven are local cafés, pubs, and fish and chip shops, places that Chatterton and Hollands (2003) would define as residual community spaces. However, as Whitehaven is now participating in the market competition for an image-conscious public, and is creating the urban infrastructure to attract them (Zukin, 1996), the gradual shift to a contemporary service economy is reflected in the arrival of a café-bar culture.

Performing Revolution Vodka Bar

Revolution Vodka Bar symbolizes current dominant change in performances of Harrogate.[2] This stylized venue is a symbol of a new café-bar drinking culture that has evolved in the town since the late 1990s. Consumers imagine and perform Harrogate Revolution through co-present practices that differ by day and by night. Revolution by day attracts a steady stream of workers and people of various ages who drop in to a modern lounge serving food and drink and vodka cocktails. By night the Harrogate 'Style Crowd' follows a route of café-bars and theme bars, where *the* place to finish the night is Revolution. Revolution is housed in the iconic Royal Baths building, a council-owned building, formerly of central importance to spa Harrogate, and now an icon in Harrogate's nightscapes.

When observing performances of Harrogate, a dominant nightscape group I term the 'Style Crowd' became evident (see Chatterton and Hollands 2003 on urban nightscapes). Members of the Style Crowd are predominantly in their twenties, work in service occupations, and are relatively affluent, white middle class residents of Harrogate and surrounding towns and cities. The creation of a café-bar scene in Harrogate in the late 1990s created a nightscape for this group that had not previously existed. A particular cultural code is at work with members reading and judging each other's appearance closely, especially clothing and demeanour. Style Crowd members are particularly keen to differentiate themselves from another nightscape community, the 'scals' or 'scags'. According to members of the Style Crowd, 'scals' wear tracksuit bottoms and trainers, drink in the Hogshead and eat at McDonalds, whereas the Style Crowd dress much more 'stylishly' and frequent stylized venues.

---

2 This paper focuses on ethnographic work carried out for my doctoral research, observing tourist and resident performances of place in Harrogate, North Yorkshire and Whitehaven, West Cumbria, England in 2003/04. To ascertain patterns in performances of eating and drinking out and opinions and perceptions of place, I followed visitors and residents, speed interviewed them, participated in eating and drinking experiences and carried out focus groups.

Figure 5.1 The Royal Baths Building, housing Revolution Vodka Bar
Source: Viv Cuthill

Figure 5.2 Service culture: Revolution Vodka Bar
Source: Viv Cuthill

A fairly fixed itinerary of Style Crowd performances operates in Harrogate on a Friday and Saturday night. A route of café-bars and theme bars leads from the centre of town, criss-crossing but avoiding the venues frequented by 'scals' and 'scags', down to the Royal Baths building where *the* place to finish the night is Revolution Vodka Bar. From 10 o'clock to midnight Revolution is packed with up to 600 people arriving at once and queuing to enter. A local loyal clientele makes of 70 per cent of Revolution's customers.

Hospitalities are highly stylized, as communicated through the service genre of Revolution Vodka Bar. In Revolution, Harrogate, there is predominance of marble surfaces, leather sofas, wood, dark red paint, and modern strip lighting on metal frames. The staff wear European-style waiter outfits, all in black, casual yet stylish. Their role is to perform the image, ambience, and style of the venue while serving colourful cocktails in chunky glasses. Their hairstyles, physique, age, and demeanour all communicate a certain cultural capital that relates directly to the image of the venue, thus illustrating the use of 'aesthetic labour' (Warhurst et al., 2000).

Revolution belongs to a national chain of drinking venues. However, Revolution's policy is not to prescribe a particular lifestyle to its customers, but to base each new bar on the character of the area, and to adapt to the customers' influence. Therefore, the design and feel of the bars is different, although the drinks philosophy and pricing is the same. I suggest that the Harrogate Style Crowd identify with Revolution, in particular, due to the design, management and performance of the hospitalities in place. Revolution has been instrumental in the diffusion and fusion of a new cultural style in Harrogate that is seen as desirable to this group of consumers, and is the central icon of their affinity group performances (Maffesoli, 1996) for as long as the venue remains in fashion.

Revolution Vodka Bar symbolizes several current issues of contestation in Harrogate, and illustrates how mobilities of fashions and tastes in eating and drinking venues alter socialities when they enter places. The desire to continue to attract a buoyant conference trade has influenced council decisions to allow many premises to convert to café-bars, thus changing the image of the staid, sober spa town into a drinking town. Performances in the streets outside Revolution by different nightscape groups have led to accusations of binge drinking and rowdy behaviour. Older residents, who appreciate the gentility of Harrogate, have stopped promenading through the town centre at night, as they feel threatened by this behaviour.

Performances of socialities and hospitalities led by the designers, managers and staff of Revolution in conjunction with Style Crowd performances has created an invisible 'style exclusion zone' within the venue, so that the service culture makes it clear to 'scals' that they are not welcome. Equally teenage respondents felt out of place in there. Finally, the quasi-privatization of the council-owned Royal Baths building, now leased to a private developer, is of concern to Harrrogate Civic Society because many local cultural societies who used to congregate and perform there in the past and are now excluded.

Therefore, the arrival of a café-bar scene is making Harrogate town centre seem less hospitable by night for certain groups of residents. An influx of stylized, contemporary, 'experiential' eating and drinking venues are creating positive images of place for some, and negative images for others. A re-enchantment of certain spaces for some groups tends to exclude others. As Harrogate town centre evolves to meet the cultural needs of affluent residents and visitors, those residents who do not exhibit appropriate cultural style find themselves on the margins. Contestation in Harrogate arises from disagreement between generations and between different social groupings about performances that are appropriate to senses of Harrogate. Economic, social and cultural change in Harrogate is partially reflected, therefore, in the socialities and hospitalities of Revolution Vodka bar.

Performing Zest Harbourside

In 2002, Zest Harbourside, owned and managed by a former London chef, Ricky Andalcio, was the first stylized café-bar to open on Whitehaven harbour. Housed in the former premises of the last wet fish shop in Whitehaven, Zest Harbourside marks a change in the eating and drinking landscapes of the working class town in transition from a post-industrial to a service economy. A link with the fishing past has been erased, and the smells of an industrial and fishing port have been replaced by the smells of glocal foodstuffs emanating from the restaurant, thus creating new tastes and odours of Whitehaven. I examine how, in performing socialities and hospitalities of place, residents and visitors perceive and experience this marginal town, and its hospitableness, and how 'tastes' of Whitehaven influence judgements of place.

Figure 5.3 Zest Harbourside – former wet fish shop
Source: Viv Cuthill

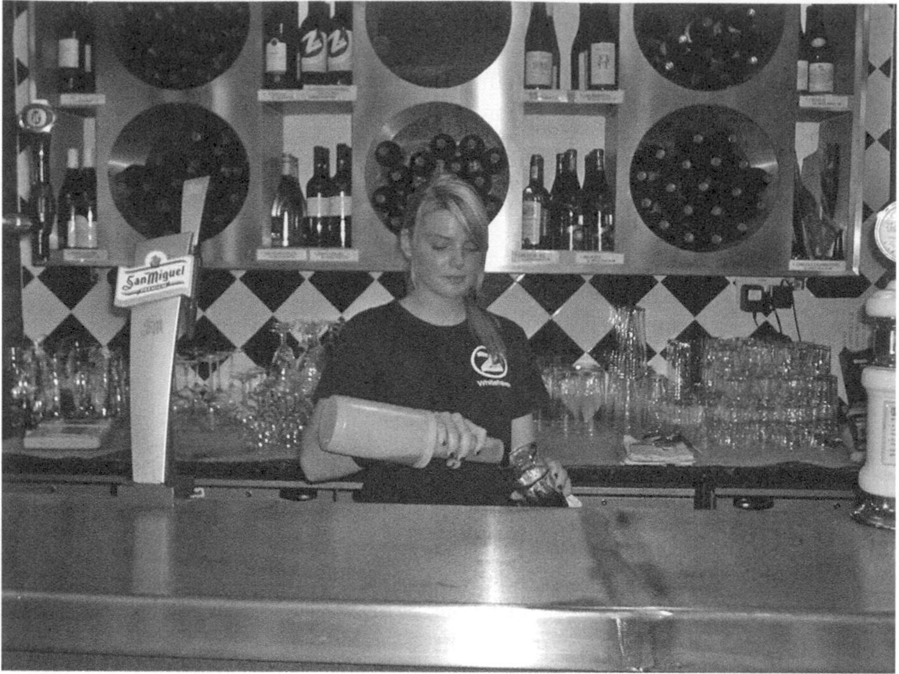

Figure 5.4 Service culture – Zest Harbourside
Source: Viv Cuthill

When Zest opened, the cream and brown painted building, canopies, logo and writing style on the exterior of the building, and external chairs and tables, clearly indicated a new kind of eating and drinking experience for Whitehaven. Originally, the internal décor was deep red walls cut across by large painted cow hide patches, with large light bulbs hanging down from wires in the ceiling; an unexpected interior design for a Whitehaven venue. Although the wall painting has since changed, the stylized effect remains. The bar area is minimal, metallic and modern, and guests can wait to be paged when a table is free in a modern upstairs lounge with leather chairs and stylish decorations.

Ricky Andalcio identified there were no quality informal dining out opportunities in Whitehaven, and created a new type of eating experience for the town. Zest is open from 11.00 to 23.00, seven days a week, the first place of its kind in Whitehaven. He describes Zest Harbourside as 'fun and funky'. The interesting feature of the menu is that all the main and side dishes are served in white or metallic bowls, there is no set meal composition, and customers choose a variety of dishes to create their own mix'n'match meal that they share from the bowls. Ricky's intention was that people could eat as much or little as they liked from this 'ultimate snacking menu'.

A wide mix of residents frequents this busy informal café-bar. Some people go there seven days a week. At any one time, it is possible to see a group of older men sharing a pot of tea and scones, a young family with children in pushchairs sharing bowls of Sunday lunch and chips, a teenage couple eating designer breakfast butties, and a middle-aged couple sharing a bowl of pan fried steak, pesto mash and a bottle of wine. The menu makes Zest very accessible and inclusive as customers can spend as little or as much as they wish at any time of the day.

Staff are all locally recruited and trained in the Zest service culture in-house, where they are obliged to 'unlearn' any previous chef or waiting experience. Ricky trains them in his service concept, which is a different style of service to typical 'Whitehaven' service. Aesthetic labour (Warhust et al., 2001) has arrived in Whitehaven, with staff adopting a European café style approach in branded black t-shirts and aprons.

Zest Harbourside marks a turning point in visitor behaviour and perceptions of eating out in Whitehaven. Whereas previously visitors walked around the harbour and moved on, without entering a café, restaurant or bar, they now sense the image and ambience of Zest, and stop off for a drink or a meal in Zest. In interviews in 2004, many cited Zest as a place they had visited and singled it out for particular praise, commenting, however, that Zest seemed completely 'out of place'. In Zest, intersections of the socialities of residents and visitors, the hospitalities of the venue and the mobilities of food fashion and style is symbolizing a new service economy that is developing in Whitehaven.

Everyday resident practices in Whitehaven are distinctive. People regularly promenade around the town and the harbour chatting, gossiping, joking, and meeting and greeting each other. One particular shopping street, King Street, is a key hub for meeting-up and passing time as clusters of people of all ages meet and chat here. As an outsider, there is a sense that everyone else is involved in an ongoing conversation that transgresses time, and that you stand out as not being from Whitehaven. De Certeau (1984) notes the mythical structure of pedestrian rhetoric as the imagination of place is enmeshed with social practices and localized street cultures are created. It is evident from resident performances that they inhabit and own the streets and public spaces of Whitehaven.

Visitors are predominantly day visitors, cyclists, walkers and maritime visitors and their distinctive attire tends to mark them out as members of particular affinity groups. They too promenade around the harbour and down King Street. As well as their physical appearance and their performances of being a visitor, gazing intently at particular landmarks, taking photographs and so on, they stand out because they are not stopping and chatting and greeting people.

The size of Whitehaven, its marginal location, and its recent economic and social history continues to bind residents into a tight-knit community. In my observations, I did not detect particular resident groups separating out and distinguishing themselves by their clothing, appearance, attitude or performative behaviours. Social groups still relate predominantly to occupation, class and family affinities and have not yet developed into lifestyle affinity groups. The place itself seems to be a social leveller, with people identifying strongly with their relationship to and pride in the town and to other people in the town, rather than to taste or style groups. When asked where they went to eat and drink, residents named every venue in town. Eating and drinking out is an activity for all and social life is determined by membership of common interest groups rather than style groups. Sociability, conviviality and maintaining social ties in place, seems to motivate eating and drinking out performances in Whitehaven, rather than expressions of status or taste.

Zest Harbourside does not symbolize issues of contestation in Whitehaven. Instead resident socialities in Zest highlight the lack of differentiation and the inclusiveness of consumption in this marginal working class town. There is no style crowd in Whitehaven creating an invisible exclusion zone around Zest. However, there are distinct disjunctures between visitor and resident senses of place.

When visitors arrive in the town they are looking for the signifiers in venues that confirm that this is 'their kind of place' (see Bell, 2004). They visually and sensually 'read' the physical environment, the appearance of people, and eating and drinking landscapes to judge whether the town is a place to be, and to locate the people and eating and drinking venues that match their particular affinities and tastes. If not, they move on. Therefore, visitors determine their own inclusion or exclusion from venues. As Pillsbury (1990) observed, specific taste communities create discontinuous geographic places through their consuming practices. Visitor socialities in Zest demonstrate that when venues provide the service culture and visual signifiers that are culturally acceptable to visitors, performances and perceptions of marginal places begin to shift.

Conclusion

Visiting eating and drinking venues is central to the sensuous and distinctive practices of performing tourist places. By examining mobilities of dominant fashionable hospitalities and socialities, we can see how people derive senses of hospitality and place from eating and drinking out.

In eating and drinking venues, service cultures convey messages of inclusion and exclusion and mark out venues as hospitality spaces for particular types of people. Hospitalities are sensed as welcoming to those who fit in, and sensed as unwelcoming by those who choose not to enter. In assessing service cultures, customers sense the hospitableness of place.

As peripheral places develop for tourist markets, they become tied to the diffusion and mobilities of innovations from cities and other cultural centres. Changing performances of socialities and hospitalities in eating and drinking venues reflect social, cultural and economic shifts in place that shape tastes in and of place over time. Both Harrogate and Whitehaven are constantly being repositioned through intersections of complex mobilities.

The place-image of Harrogate as a traditional spa town offering a particular type of quality eating and drinking experience is challenged by change in the built environment that is reflected in service cultures and in the performances of visitors and residents. In this highly developed place of consumption, segmented affinity groups create temporary communities based on expressions of lifestyle and taste, and exclude others in the performances of their lifestyle. Identities are expressed through the performances of sociality in particular hospitality spaces, such as Revolution Vodka Bar.

Frequenting stylized venues may be associated with placing taste, identifying with the style crowd and accruing cultural capital, however, the effects of the aestheticization and stylization of eating and drinking venues in geographically and culturally peripheral places cannot be categorized simply. The ambience of Whitehaven, just as in Harrogate, is created through the interplay of venues, people and performances in public spaces, creating the 'aura' that gives places a unique colouring and odour (Maffesoli, 1996: 22). The development of a tourism sector is beginning to alter the built environment.

Service cultures and consumer performances. However, there are disjunctures between resident and visitor senses of hospitality in place. Resident performances in eating and drinking venues, such as Zest Harbourside, are inclusive, community-based and not segmented. Visitors to Whitehaven, who express their affinities through segmented performances of consumption, only eat and drink in places they sense as culturally acceptable, such as Zest Harbourside, a venue they describe, however, as 'out of place'.

Despite the fact that Revolution Vodka Bar is a chain venue, and Zest Harbourside is an individually owned operation, the design of the service genre is remarkably similar in terms of material culture and staff hospitalities. However, in taking into account the performances of customers and their socialities as part of the service culture, different senses of place and of the hospitableness of place are evident.

The cultural acceptability of place is, therefore, partially sensed in the hospitalities and socialities performed in eating and drinking venues. Stylized venues re-position and re-rank places within global cultural circuits. The hospitableness of commercial hospitality spaces is constantly negotiated and renegotiated in the intersections of performances of affinity groups and service staff in specific physical and material environments. Venues represent dominant inclusions and exclusions in relation to global fashions and tastes in commercial hospitalities and place.

References

Baerenholdt, J.O., Haldrup, M., Larsen, J. and Urry, J. (2004), *Performing Tourist Places* (Aldershot: Ashgate Publishing).

Bell, D. (2004), 'Taste and Space: Eating Out in the City Today' in Sloan (eds).

—— (2006), 'Hospitality and Urban Regeneration' in Lashley, Lynch and Morrison (eds).

Bell, D. and Valentine, G. (1997), *Consuming Geographies: We Are Where We Eat* (London: Routledge Books).

Bourdieu, P. (1984), *Distinction: A Social Critique of the Judgement of Taste* (London: Routledge Books).

Chatterton, P. and Hollands, R. (2003), *Urban Nightscapes: Youth Cultures, Pleasure Spaces and Corporate Power* (London: Routledge Books).

Crang, P. (1997), 'Performing the Tourist Product' in Rojek and Urry (eds).

Crouch, D. and Desforges, L. (2003), 'The Sensuous in the Tourist Encounter', *Journal of Tourist Studies*, **3**(1), 5–22.

Crouch, D., ed. (1999), *Leisure/Tourism Geographies* (London: Routledge Books).
De Certeau, M. (1984), *The Practice of Everyday Life* (Berkeley, CA: University of California Press Books).
Derrida, J. (2002), *Acts of Religion* (London: Routledge Books).
Finkelstein, J. (1989), *Dining Out* (Oxford: Basil Blackwell Books).
Goffman, E. (1956), *The Presentation of the Self in Everyday Life* (London: Penguin Books).
King, A.D., ed. (1996), *Representing the City: Ethnicity, Capital and Culture in the Twenty First Century Metropolis* (London: Macmillan Publishing).
Lashley, C., Lynch, P. and Morrison, A., eds (2006), *Hospitality: A Social Lens* (Oxford: Elsevier Books).
Laurier, E. and Philo, C. (2005), *The Cappuccino Community: Cafés and Civic Life in the Contemporary City – Field Report 3: What Café Customers Do* (Department of Geography, University of Glasgow) [website], http://web.ges.gla.ac.uk/~elaurier/cafesite/texts/customer.pdf >, (accessed 20.08.06).
Maffesoli, M. (1996), *The Time of the Tribes. The Decline of Individualism in Mass Society* (London: Sage).
Pillsbury, R. (1990), *From Boarding House to Bistro: The American Restaurant Then and Now* (London: Unwin-Hyman Books).
Pine, B.J. and Gilmore, J.H. (1999), *The Experience Economy: Work is Theatre & Every Business a Stage* (Boston, MA: Harvard Business School Publishing).
Richardson, P. (2000), *Cornucopia: A Gastronomic Tour of Britain* (London: Abacus).
Rojek, C. and Urry, J., eds (1997), *Touring Cultures* (London: Routledge Books).
Sheller, M. and Urry, J., eds (2004), *Tourism Mobilities* (London: Routledge Books).
Sloan, D., ed. (2004), *Culinary Taste: Consumer Behaviour in the International Restaurant Sector* (Oxford: Elsevier Books/Butterworth-Heinemann).
Telfer, E. (1996), *Food for Thought: Philosophy and Food* (London: Routledge Books).
Urry, J. (2002), *The Tourist Gaze*, 2nd ed. (London: Sage).
Warde, A. and Martens, L. (2000), *Eating Out: Social Differentiation, Consumption and Pleasure* (Cambridge: Cambridge University Press).
Warhurst, C., Nickson, D., Witz, A. and Cullen, A.M. (2000), 'Aesthetic Labour in Interactive Service Work: Some Case Study Evidence from the "New" Glasgow', *The Service Industries Journal*, **20**(3), 1–18.
Zukin, S. (1995), *The Cultures of Cities* (Oxford: Blackwell Books).
—— (1996), 'Space and Symbols in and Age of Decline' in King (eds).
—— (1998), 'Urban Lifestyles: Diversity and Standardisation in Spaces of Consumption', *Urban Studies*, **35**, 5–6, 825–839.

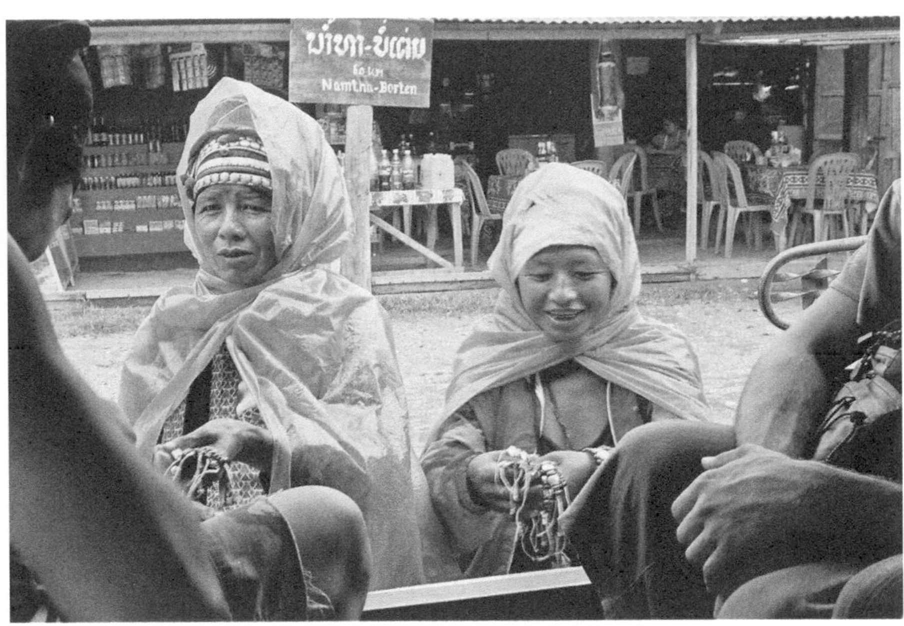

Opium Sellers **(Luang Namtha, Laos, 2004) by Elly Clarke**

Chapter 6

Hospitality, Kinesthesis and Health: Swedish Spas and the Market for Well-Being[1]

Tom O'Dell

In 1998, one of Stockholm's historic bathhouses produced a book entitled, *Spa: behandling, träning, mat, avspänning* (Spa: treatment, training, food, relaxation [Swanberg, 1998]). Between the covers of the book the reader could find a wealth of chapters describing exotic spas, invigorating massages, 'living muds' and treatments affectionately referred to as 'egocentric' (Swanberg, 1998: 19ff.). More than an ode to the world of spas, this was a publication with a problem to tackle. As the book's editor explained:

> Most people in Sweden are unaccustomed to enjoying and accepting something without a thought of reciprocation… It is not even fully acceptable to spoil oneself by letting another person, with soft massaging hands, dedicate an hour at a time to such a humble little part of the body as the foot… it's a shame to hold back (Swanberg, 1998: 10).[2]

At issue here was the packaging and promotion of a 'new' product (with old roots) that many Swedes viewed with unease. Part of the problem confronting Swedes was the question of whether or not one could morally defend the purchase of this type of personal service.

At the time, Swedes had just become embroiled in a heated public debate over the appropriateness of paying for private services such as house cleaning – a dispute that continues to this day and is popularly referred to as 'the maid debate'. The problem here was that in Sweden, the commercialization of home services raised uneasy questions, at least in the eyes of some, about growing class divisions, and the return of a class society that Sweden had seemingly left behind long ago.[3] In

---

1 The research presented here has received support from the Swedish Research Council. I also want to thank the participants in the 'Mobilizing Hospitality: The Ethics of Social Relations in a Mobile World' workshop held at Lancaster University, 26 September and 27, 2005 for their comments on an earlier draft of this paper.

2 All quotes which were originally in Swedish have been translated by the author.

3 Most recently, the maid debate came to the fore of public attention in the wake of the 2006 national elections in which the conservative coalition came to power promising, among other things, tax reductions that would make cleaning services more affordable to private families.

this context, the commodification of something as private and personal as a caress seemed to push these issues a step further.

This being said, however, it could also be argued that at the heart of the problem lay a series of questions concerning processes of mobility and their role in a context of commercialized hospitality. The book, *Spa*, seemingly portrayed a highly democratic world open to everyone, but what sorts of people would one actually find circulating within the perimeters of the spa? And while many spas promised to help their guests find new sources of energy locked up inside their bodies, one was left to wonder how a spa could set these powers in motion? And did a visit to a spa represent a physical and symbolic movement into a new future – delineated by new understandings of 'the road to better health' – or did it perhaps represent a step in the direction of an increasingly segregated class-bound society of masters and servants?

Bearing these questions in mind, the following chapter focuses upon three Swedish spas and the role the management of micro-processes of mobility have played in these facilities' attempts to open a space of hospitality that is capable of producing a sense of well-being within them. While most of the contributions to this volume have highlighted larger forms of mobility linked to such phenomena as migration and tourism, this text directs its attention to a more corporeally anchored analysis of the relation between mobility, space and hospitality. In opening, historic materials are used to help illuminate the shifting significance that has been attached to the notion of hospitality in times past, and to problematize how an increasingly commercialized atmosphere of hospitality came to be organized through culturally and spatially defined regimes of mobility that controlled the circulation of bodies in nineteenth-century Swedish spas. Following this, the discussion then turns to present day Sweden and the manner in which spas use processes of mobility management today to affect their guests' senses, and thereby, their experience of being cared for. In closing, the text draws attention to the effects regimes of mobility have upon those working in these spaces of hospitality, as well as guests' perceptions of them. Throughout it all, the text endeavours to argue for a need to more closely examine the implications the management of bodily movements has had for the cultural orchestration of hospitality, particularly as it has come to expression in commercialized settings.

Hospitality, Obligation and Circulation

Before proceeding, however, it should be acknowledged that writing about hospitality in relation to spas positions the subject in a rather special context that is worth briefly reflecting upon.[4]

---

4 It should be noted that the form of hospitality interrogated in this chapter is characterized by a situation in which the guest/host relationship is bound by commercialized processes of exchange. It is, in other words, a phenomenon limited and controlled by contextually defined laws (in the plural) that place obligations upon both the guest and the host. As a result, it never approaches the phenomenon that Derrida called 'absolute hospitality' (Derrida, 2000: 83). Nonetheless, the immersion of the notion of hospitality in a discourse of exchange and market economics has had strong implications for the manner in which hospitality has been

In many ways, it might be argued that the offering and extension of hospitality was, from the very beginning, a phenomenon centrally concerned with issues of both health and mobility. Prior to the spread and institutionalization of commercial forms of accommodation, travel was a precarious and at times dangerous activity. The extension of private forms of hospitality that included sustenance and safe shelter was all but a prerequisite for its conduct (cf. King, 1995: 221ff.). But to the extent that this is true, the laws of hospitality have long been limited, controlled, debated and even subjected to political and personal interest.

In early modern England (1400–1700) for example, hospitality was perceived to be a Christian duty regulated by values and norms of behaviour dating back to classical Rome. However, the exact implications of what the duties of hospitality included were contested (Heal, 1990; Selwyn, 2000: 21). For example, it was generally agreed that the good host had an obligation to extend his/her hospitality to the poor. Likewise, the landed elite were expected to open their households to foreigners and strangers, but the norms of the day also urged temperance and moderation. The needy (whether a neighbour or a stranger), some argued, should be cared for, but not overindulged. Unfortunately, establishing the status of a stranger was not always an easy task. Thus, while some moralists argued for the merits of restraint, others pointed to the biblical lesson of Lot, and reminded their contemporaries of the fact that even unassuming strangers could be angels in disguise, worthy of care and hospitality (Heal, 1984).

Complicating matters even more, however, some critics of the time lamented what they perceived to be the decline of English hospitality (cf. Walton, 2000: 58), and the fact that in reality when hospitality was extended to the poor, it all too often implied little more than the doling out of table scraps and leftovers to those who were banished to the gatehouse (Heal, 1984: 75ff.). Instead of taking the form of an unselfish Christian duty, hospitality, it was pointed out, was increasingly directed towards peers and people of standing, taking forms that were often intended to endow the host with increased social status, honour and political influence. In the eyes of some moralists, this raised more than a few uncomfortable questions as to whether or not this manner of giving for gain was an appropriate Christian activity (Heal, 1990).

Private hospitality, in other words, was an important means of offering help, care and provisions to strangers as well as disenfranchised members of the local community; however, it could also be used strategically to further a host's own ulterior social, political and economic objectives – and in this sense possessed qualities that many associate with commercialized forms of hospitality today (cf. Telfer, 2000: 40ff.). It was, in short, a morally laden social field of exchange and interaction whose bounds and limits were continuously contested and debated.

Over the course of the eighteenth and nineteenth centuries, commercial forms of hospitality came to increasingly replace many of the functions that were once tied to private hospitality. But this was a successive and slow process in which it is possible

---

popularly perceived, consumed, produced and spatially organized. The focus of analysis for this text lies, in other words, upon a very conditional context of hospitality, and it is the issue of conditionality (and its consequences) which I argue is in need of further interrogation.

to discern the parallel existence of older morals and norms alongside newer, more modern market-driven concerns. These processes came to particular expression in a wide range of European spa settings that united new forms of commercial leisure and secularized notions of the power of medicine (and science) with a series of older moralities and beliefs (cf. Bacon, 1997; Mackaman, 1998).

In Sweden, the very first spas were established in the last decades of the seventeenth century, and their popularity amongst the Swedish royalty, aristocracy and cultural elite played a central role in facilitating the spread and development of new facilities in the eighteenth century (Mansén, 2001). By the mid-nineteenth century, however, many spas were forced of economic necessity to recognize the role middle-class patrons could play as paying patrons, and quickly developed a wide array of special mid-priced package deals for these prospective guests.

The complexity of the situation can in many ways be exemplified by *Ramlösa Hälsobrunn*, a popular spa located near Helsingborg on the Southern Swedish coast. As it turns out, *Ramlösa Häslobrunn* was fortuitous enough to win the favour of the King and Royal family of Sweden who repeatedly visited the facility, in the early nineteenth century. This in turn attracted the aristocracy of Scania, many of whom built their own summer homes around the spa (Mansén, 2001). All of this bode well for *Ramlösa Hälsobrunn*, but problems still lie on the horizon, for while these visitors made private contributions to the spa, and even organized balls, concerts and theatre performances – the proceeds from which were donated to the spa (*Ramlösa Hallået* 1899, 5; 1912, 7) – these contributions were never substantial enough to assure the economic security of *Ramlösa Hälsobrunn*. Indeed, they usually did not account for more than a sixth of the spa's annual income (Mansén, 2001: 86). In order to make ends meet, the private owners of the spa had to endeavour to attract the middle and lower classes, and they did so by building special ballrooms and living quarters for them, as well as by organizing balls, fireworks and carnivals. Management at the resort even petitioned the King for the rights to open a casino, and although permission was never granted, they nonetheless opened a highly popular gambling hall that not only attracted the spa's seasonal guests, but even drew Danes from across the sound who flocked to the facility on Sundays to place their bets (Lindblad, 1907: 35ff.).

However, these and other efforts to continuously attract new groups of guests to *Ramlösa Hälsobrunn* were not enough to prevent the facility from repeatedly facing economic problems that prompted regular shifts in ownership. Despite these pressures, the spa, nonetheless, still made efforts to extend its hospitality to the poor. In part, a policy of openness and tolerance was prescribed by older patriarchal traditions and senses of responsibility. In Sweden, for example, the taking of water was indelibly defined as an activity open to everyone, from the wealthiest to the most destitute members of society.[5] And this remained the case at *Ramlösa Hälsobrunn* throughout the nineteenth century (Mansén, 2001: 505). However, the extension of hospitality to the disenfranchised did not end there as it perhaps could have in light

5 See Strang (2004) for a discussion of the significance that has been attributed to water in a wider European context.

of the economic difficulties and climate of growing economic rationality that faced the spa.

Instead, efforts continued to be made to keep *Ramlösa Hälsobrunn* open as a 'hospitable space' (Dikeç, 2002: 228) that even had room for those in need. On average, the spa received approximately one thousand visitors per season in the early and mid-nineteenth century. But in a year such as 1835, 402 of these visitors received some form of economic support from the spa for their stay.[6] Thus, while *Ramlösa Hälsobrunn* attracted the upper echelons of the social and economic elite in Southern Sweden, it also remained a rather heterogeneous environment in which very different groups of guests, visitors and staff (including maids, bath maids, cooks and attendants who remained as inconspicuous as possible) moved around in proximity to one another, and in which the moral atmosphere of the times still prescribed a degree of tolerance on the part of guests that made this form of hospitality possible.

However, the emphasis of analysis should perhaps be placed upon the issues of proximity rather than tolerance. For while the socio-economic mix of guests at *Ramlösa Hälsobrunn* was diverse, guests from different social strata were by and larger strangers to one another, and remained so throughout the course of their stay. In this context, the routines, rules and physical organization of the spa helped to minimize contact between different categories of spa guests. The taking of water, for example, was an activity in which most guests participated, but it was organized around a time schedule in which the poorest patients arrived at the spa's well early in the morning (around 04.00) while guests of higher and higher standing arrived ever later in the day. In a similar manner, the existence of first and second class bath houses, living and medical facilities for the poor, special ballrooms and dance locations for the peasantry, privately owned summer houses for the wealthy, and finer ballrooms and reading rooms for the cultural elite, all contributed to a pattern of perpetual motion in which guests circulated and socialized within circles constituted by patrons of similar standing. In this sense, the ability to maintain an assemblage of steering mechanisms, a 'social smoothing machine' of sorts (Bogard (2000: 272ff) after Deleuze and Guattari (1987: 474ff)), that kept different categories of guests separated even as they moved around in proximity to one another, was integral to the production of a hospitable space at *Ramlösa Hälsobrunn.*

This is important because while pure hospitality necessitates a condition of unconditionality (Derrida, 2000: 25), most forms of hospitality as we know and experience them, are conditioned by culturally bound codes and laws. In this context (of conditional hospitality) the production of hospitable space not only requires degrees of tolerance, but even, as Mustafa Dikeç has argued, acceptance and recognition (2002, 236). Unfortunately, cultural atmospheres that perpetuate attitudes of acceptance towards others can be fragile. Hospitable space is not something that can simply be created once and for all; it has to be opened and reopened, continuously made and remade. The case may be that this not only requires mechanisms that

---

6 This help often took the form of smaller subventions covering the costs of the baths, food, housing, or medicines of those without means, but were later consolidated into a limited number of full-season cures (five to six weeks) for those deemed to be in most need (Lindblad, 1907: 49).

facilitate attitudes of acceptance, but perhaps just as importantly, cultural practices that diminish tension and friction, and that thereby facilitate the recognition of alternative life circumstances by reducing the chances that people close the door of recognition to one another. At *Ramlösa Hälsobrunn* the practices of corporeal movement that existed there worked in this way, for a while, contributing to the perpetuation of a space of hospitality that allowed for degrees of mutual recognition and acceptance between groups of guests.

But as a site of conditional hospitality, *Ramlösa Hälsobrunn* always remained a rather ambivalent space of hospitality. It was a place in which segments of guests sought to raise funds for the inclusion of others, but often did so through social activities whose fees prevented the participation of those others. And these activities were conducted in a context in which 'the benevolent' paid for the (first class) right to nurture their own well-being without the intrusion of others. From the perspective of the spa owners, the morals of the day made it difficult to justify the full and automatic exclusion of any group of sickly patrons on solely economic grounds, but economic realities also placed constraints on the degree to which the spa's hospitality could be extended to those without sufficient means, while simultaneously reinforcing the importance attached to both the paying customer, and her/his potential charitable goodwill. In the face of these tensions, the production of well-being (or at least the sensations of well-being) was intimately linked to the careful orchestration of each guest's mobility, and guests themselves (particularly those of standing) participated actively – even taking the initiative at times – in the perpetuation of these processes of control and segregation.

As Roy Wood has argued (1994), actors in the hospitality industry still work to control the movements of their guests and the interactions between them. Unfortunately, little systematic attention has been paid to these mobility bound processes, and the significance these have in relation to commercialized forms of hospitality. Instead, as other commentators have noted (Brotherton, 1999: 168) efforts made by management theorists, particularly those working in the field of hospitality management, to problematize hospitality as a phenomenon have all too often fallen into the trap of either striving to delineate the contours of an industry, or skewing the focus of the discussion in the direction of the products delivered: food, beverages, accommodation and entertainment. This is not to say that hospitality management researchers have completely failed to acknowledge the guest/host relationship, or the fact that hospitality implicates processes of mobility – at least to the extent that some scholars have recognized that processes linked to issues of arrival and departure are important to understanding hospitality (see for example: King, 1995; Mullins, 2001: 14) – but that these discussions have generally tended to be highly instrumental in nature. Rather than problematizing hospitality, they have all too often, wound up offering models and methods for its 'deliverance'. Consequently, hospitality has been readily aligned with the provisioning of services, which are themselves fetishized as commodities with a definable exchange and use value that can be optimized. In the process, hospitality seemingly becomes universally and democratically accessible as a market item. Questions which are comfortably avoided, and which I am trying to draw attention to here, concern tolerance and the mechanisms by which commercialized institutions of hospitality, such as spas, affect, sort and categorize strangers, defining some as guests, and some as employees, while disregarding those who are not present as irrelevant.

But this discussion can be taken even one step further. For example, if we move our focus of analysis from the nineteenth century into the years around the new millennium, we find a cultural context in which Swedish spas in particular, but even the spa industry as a whole, increasingly frame their hospitality in terms of the personal experience of well-being that they can produce (cf. O'Dell, 2005, 2006).[7] That is, where the nineteenth-century Swedish spa offered hospitality in the form of a welcoming and warm *social environment* in which guests could dance, attend concerts and dream of chance encounters in the park, the space of hospitality at spas today is organized around more *individually oriented experiences* in the form of a slower pace, a time of reflection and the potential to find new energies and strength. Rather than promoting themselves as arenas of social interaction, Swedish spas now market themselves as places that produce experiences of well-being that emanate form within – albeit with the aid of others, in a specific material environment, and through the orchestration of new modes of mobility. There is, in other words, a linkage here between current interests in personal health, the consumption and production of experiences (cf. Pine and Gilmore, 1999; O'Dell and Billing, 2005), the offerings of the hospitality industry, and issues of mobility that remain largely unexplored (although see Thrift, 2000). As a means of illuminating some of the ways in which these issues and the questions outlined above have become entangled in spas today, let me turn to two Swedish spas: *Hasseludden* and *Varberg*.[8]

Well-being, Affect and the Cultural Organization of Micro-mobilities

Varberg and *Hasseludden* are interesting because while both facilities are in the market of selling the experience of 'well-being' to primarily Swedish consumers, they have chosen to do so in very different ways. In part, these differences have a historical explanation. Neither of these two spas was originally constructed to function as a spa. *Varberg*, located just south of Gothenberg on the western coast of Sweden, was a sanatorium that was later used as a health resort and converted into a conference center in the 1990s. Echoing its past, the facility's main building clearly reflects the early modern aesthetics found in most hospitals built in the early part of the twentieth century. The main lobby, however, takes its cues from English aesthetics and includes walls of panelled wood and deep-green, stuffed leather furniture. This is a dark lobby by Scandinavian standards. But together, the spa's historic roots, twentieth century institutional architecture and classical English-style

---

7 While the linkages to the field of medicine are stronger in other European contexts (cf. Weisz, 2001), a quick survey of any newsstand witnesses to the fact that the growing interest in the experiential aspect of the spa visit is also on the rise in many of these settings.

8 The following section is based upon fieldwork I conducted at *Hasseludden*, *Varberg* and other spas. It is constituted out of a form of autoethnography based upon my own impressions of the sites in which I worked. But it is also informed by interviews I have conducted with spa management, employees and patrons. For a discussion of the advantages and limits of autoethnography, see Marcus, 1998: 246 and Reed-Danahay, 1997.

lobby all work to symbolically assure visitors that this is an establishment rich in tradition, with a well-rooted heritage of its own that is coupled to a deeply anchored history of professionalism.

In contrast, *Hasseludden*, located in the Stockholm area, was built in the early 1970s by the Swedish Labor Union, Lo, with the intention of using the facility as a conference and educational center. The building was designed by Yoji Kasajimi, and incorporates Japanese styles and aesthetics with the functionalistic ideals that dominated much of Swedish architecture in the early 1970s. In the 1990s, the Japanese theme permeating the building's architecture was expanded upon and developed into the *Yasuragi* spa concept.[9] This concept includes Japanese inspired treatments, massages, bathing facilities and several restaurants serving Japanese food. Guests are even requested to wear traditional cotton Japanese robes called *yukata* throughout the length of their stay.

Despite these differences in style and profile, however, both spas share the ambition of offering their patrons access to a space of hospitality whose contours they themselves define in terms of such sensations as 'harmony', 'relaxation' and 'well-being' (see *Varbergs Kurort Hotell & Spa Brochure* 2004; and *Sinnesro* 2002). In order to accomplish this, they have a wide range of props and strategies at their disposal that are directed towards the management of the smallest of movements (and their effects upon the senses). For example, the entrance lobby of *Varberg*, and the Japanese theme of *Hasseludden*, both work as framing mechanisms that help to signify that these are places that force altered states upon those moving about within their confines. The lobby at *Varberg* evokes, as a manager at the spa explained to me, a 'sense of serenity' amongst visitors. As people enter the lobby in the morning, they not only move into a room that is darker than the outside surroundings, but they actually do enter an institutional environment (that of a previous sanatorium). Loud voices and noises produce echoes that, like ghosts, bounce off the walls and hard floors, and come back to haunt all within the lobby. This does not only lead to the dampening of voices, but even (at least for some) diminished and more constrained gesticulations, and softer foots steps, which in their turn lead to slower and shorter strides. In this sense, synesthesia, 'the transposition of sensory images or sensory attributes from one modality to another' (Marks, 1978: 8 quoted in Feld, 2005: 181; see Howe, 2005: 292) is an important catalyst in the production of a 'sense of serenity' in the lobby of *Varberg*, that implicates the collusive interaction of the eyes and ears upon the unsuspecting musculature of the body.

The Japanese theme at *Hasseludden* – featuring a sparsely decorated interior – plays upon the notion of stripping away all that is excessive, of returning to a purer aesthetic and spiritual state. This aesthetic state is then transferred into the body of visitors with the aid of the *yukata* robes that guests are requested to wear. As one of the managers of *Hasseludden* explained to me, the *yukata* sit tight around the body and inhibit movement. They are, in other words, physically constraining, and force visitors into a new slower rhythm of movement, which is then further accentuated

9 *Hasseludden* explains that Yasuragi means 'harmony', or 'inner calm'. For many years, it was also the name they gave to their bathing facilities (*Sinnesro* 2002, 5), although those facilities are now called, 'Kyarabi', which means 'gracious beauty' (www.yasuragi.se/).

by thin, light cotton slippers which guests are provided along with their *yukata* and encouraged to wear. The slippers do not sit well on the foot, and have a tendency to glide off as you walk. They are also slippery on the sleek wooden and stone floors of *Hasseludden*, providing a sensation akin to ice-skating with untied ice skates. The combined result of the *yukata* and slippers is a radically altered kinesthetic condition provoking not only new schemata of movement (shorter steps, shuffling feet, the working of the toes to keep the slippers on, and so on), but even a new form and focus of concentration (and consciousness) upon bodily tempo, tactility and coordination.

However, the efforts of spas to reach into and affect bodies do not stop there. Massage and treatment rooms are also carefully arranged and organized. As part of this, different colours are used in the interiors of these rooms in order to evoke different moods. At *Varberg*, for example, the spa management explained that they periodically changed the colour schemes of the waiting rooms that lie in conjunction with the Tong baths, in the hope that this would provide returning guests with the sensation of having experienced something new. In the world of hotels, rooms are continuously refurbished as they age and become worn. Refurbishing is, in this case, performed in order to maintain a certain standard of convenience. In contrast, the logic driving the refurbishment of the Tong Bath waiting rooms is more explicitly driven by a desire to reach into the visitors' being and have an emotive affect upon him or her.

The boundaries of rooms are in this sense not only delineated by doors, walls and windows, but also by divergent atmospheres of sensuous stimulation. Rooms containing indoor pools, for example, feel humid, smell of chlorine and possess a unique reverberative sound quality that is very different from a carpeted hotel corridor. Similar principles work as one moves from room to room in a spa. One area may smell of fresh flowers, while another is marked by a new and strange odour that may be an aloe vera oil, or a burning incense. The result is that as visitors dampen their voices, and physically slow down, their olfactory senses are constantly triggered, put in motion, and spurred to new states of arousal. Continuously, working hand in hand in this way, competing forms of micro-mobilities produce a series of sensations in guests which they recognize, define and identify in terms of 'well-being'.

This sensation of well-being is made possible because spas have a materiality that not only shapes the experiences of people who move about within their confines, but this materiality 'matters' (Miller, 1998) because it has a way of mapping itself into the bodies of patrons. By stimulating the senses it creates impressions and emotions, and informs consciousness – potentially affecting visitors in ways that may not always completely be apparent to them. Understanding the role of the body and the senses in this context is important because, while our senses play an enormously important role in shaping conscious thought, it has become increasingly clear that a very large portion (according to some, over 95 per cent) of the impulses and information that we process work below the radar of conscious thought (Lakoff and Johnson, 1999: 13; Thrift, 2000: 40). As Tor Nørretranders has pointed out, 'The bandwidth of language is far lower than the bandwidth of sensation. Most of what we know about the world we can never tell each other' (1998, 309).

The spa, in this context, is not only a geography of health and leisure consumed by patrons, but through its material organization, it opens a space of commercialized and market-driven hospitality that works the senses, stimulating different forms and tempos of mobility in its attempt to affect the moods and emotions of patrons. To the extent that it succeeds in making visitors feel reinvigorated, it does so by a rich multiplicity of sensuous cues which are only partially registered consciously in the minds of guests. It is the 'feel' of the hospitality experience that is more important here than any logical understanding and awareness of it.

In order to succeed, spa managers invest much effort in planning, designing and maintaining their facilities. However, the space of the spa is not simply something that is designed and orchestrated from above; it is also created as people move through it and produce competing understandings of it. Let me expand upon this point with the aid of two more examples, before proceeding to some final comments.

Orchestration and Mobile Bodies

Spas such as *Varberg* and *Hasseludden* work hard at creating an image of themselves as hospitable spaces of calm, harmony and well-being. But when viewed from slightly different perspective, they can be seen as sites of intensive activity. *Varberg*, for example, employs over 200 people (predominantly women) in any given month. This is a workforce that is constantly in motion in the spa, silently and discretely coming and going from room to room. The majority of these employees work on a part-time basis, and like most employees in the service sector, receive modest wages. Many of those who work within the industry find their jobs so physically demanding that they cannot work full-time. As one branch publication explained, 'Spas give you the impression of being about luxury, but no one in the field is strong enough to work full-time. At the same time, they have to make a living, so they keep on working' (Kellner, 2003: 22). The ability of masseuses to plan their free time and recuperate from a week of heavy work is hindered by a rotating system used by some spas, in which personnel are on-call and expected to work when they are needed. Weekends tend to be the busiest time of the week for spas and as a result it is not uncommon for employees who are on-call to find themselves working weekends on which they would otherwise be free. Other strategies are also available. Another one of Sweden's larger spas relies heavily upon young women who come directly from diverse masseuse training programs and puts them to work full-time. As a result many quit their jobs within a year (*Svensk Hotellrevy* 2003, 23). Here it seems to be easier to replace worn-out employees than to make their working routines bearable.

Spas are labyrinths of mobile bodies. In the worst of cases they are sites of stress production invoking a revolving-door policy that consumes bodies, wears them out and ultimately discards them. In other cases, attempts are made to avoid this by employing people on a part-time and flexible basis, or by encouraging employees engaged in strenuous activities to perform less physically demanding tasks from receptionist duties, to serving in the facilities' restaurants. Work schedules divided into days, half-days, weekends and evenings, steer the constant flow of bodies between employment positions and stations.

This is, however, not a flow that is limited to the bodies of employees. Spas such as *Varberg* and *Hasseludden* tend to accommodate thousands of visitors every week. No single service could meet the wishes of all these clients. Consequently, larger spas such as these have had to develop innovative strategies for delivering mass services that seem to be entirely individualized. One of the organizational techniques that makes this possible at *Varberg* is a timetable called an 'Activities Menu'. The menu divides each day into blocks of time ranging from 30 to 60 minutes, and guarantees exactly which activity or treatment patrons can participate in at any given time on any day. A sign-up sheet posted in the spa's lobby specifies the number of people who can participate in each activity. No two days offer exactly the same schedule, and the rule of first come, first serve, applies to many of the activities on the sign-up sheet. While relaxation is the goal, the clock rules here in much the same way it has throughout the industrial era.

Indeed, time is of the essence, so *Varberg* instructs guests to arrive at their pre-ordered treatments five minutes ahead of schedule. By spa standards, these are rather lax time constraints. Other spas go so far as to demand your presence 15 minutes in advance of your massage or treatment and warn that lateness will be deducted from the time of your treatment. The disciplined participation of both the service provider and the spa patron are, in other words, a precondition of hospitality in this context. Both share the responsibility of watching the clock and meeting time requirements. The trick for the spa is to subtly teach its clientele how to 'properly' partake of its hospitality – teaching them the production rules of being served, 'spa-style' – without appearing to be domineering, and without detrimentally affecting the appearance that the spa is extending a very personalized form of hospitality to its guests.

At *Varberg*, work schedules, and activities menus are crucial organizational instruments that help the spa to handle and facilitate a regime of flexibility that keeps both patrons and employees on the go – moving them from room to room, providing them with a series of activities to keep them busy, and clearly defining the stations and time frames in which to perform their work (or leisure). Other cultural theorists have assertively argued for the need to more greatly appreciate the capacity of flexibility to 'arouse anxiety' (Sennett, 1998: 9) and 'uncertainty' (Bauman, 2000: 147) within the ranks of the labour market. These are aspects of working-place flexibility that spa employees face; somewhat paradoxically, however, it is also via an exposure to similar processes and organizational strategies that spa patrons strive to flee stress and anxiety – and in many cases claim to succeed in this pursuit. Regimes of flexibility, it seems, do more than produce stress.

In other places (O'Dell, 2004) I have argued for a need to better understand the kinesthetic tension between cultural processes of stasis and mobility and the manner in which they become morally charged in daily life. By juxtaposing the work schedules of spa employees with the activity schedules offered to visitors, my intention here is to argue for a need to further interrogate the question of what the demands (and expectations) of flexibility and mobility do with us in shifting contexts. Why do regimes of flexibility and mobility seem to 'break people down' in some contexts, but 'build them up' in others that come to be culturally defined as 'hospitable'? As I have been arguing throughout this text, spas endeavour to produce a feeling of well-being by carefully managing processes of mobility and kinesthesis, which target the bodies and senses of their guests. Interestingly, at *Varberg*, corporal tempos of activity, which among other things, are associated with productivity in working life, are re-invoked and re-contextualized in a manner that (perhaps not so surprisingly) reassures guests of the facility's ability to have a productive and rejuvenating affect upon them. This may seem paradoxical, even 'illogical', but to the extent that it works, it does so because it all makes sense to the body, at some level – speaking to it, if you will, through a well rehearsed and internalized language of flexibility and production that it has come to readily understand. And perhaps, that is what is most important in this context.

As Nigel Thrift has pointed out (2004), we live in a world in which our consciousness lags after us by 0.8–1.5 seconds. That is, most people require nearly a half to a full second of 'thinking time' in which to become aware of occurrences around them. Consciousness is, in other words, something that forms and develops in us, and because of this, it is something of an after-the-fact construction (Nørretranders, 1998: 289ff.). In targeting our senses, spas are essentially engaged in a project that aims at colonizing the pre-reflexive gap between our bodies and our consciousness, working to affect the latter through the former. The sensation of 'feeling better' in this context is produced on one level through the corporeal impressions that are generated out of the spatial practices encompassing guests and staff alike.

But these sensations are reinforced through very particular regimes of cultural organization that keep people moving in ways that minimize unwanted distractions, and interruptions, that could otherwise disturb the overall impression of the spa experience. At *Ramlösa Hälsobrunn*, very different groups of guests were ever present in the nineteenth century, but organized in a manner that diminished the degree of their interaction. And while they may not have socialized much together, they did at least share a limited degree of mutual recognition of one another. Today, the market mechanism of pricing works to diminish the diversity of the guest population at Swedish spas,[10] as does the promotional imagery that they use to attract customers. However, the line of difference that exists between staff and guests is still a point of potential tension that spas must contend with. The use of uniforms, professional language, 'staff only' areas of access and ritualized means of beginning and ending treatments are all used to help create and maintain the 'right atmosphere' – for the reception of this particular form of commercialized hospitality – but they do so by disguising class differences, economic inequalities, hierarchies of servitude, the stress of work schedules and the potential physical frailties of staff members. All the while, the time schedules discussed above keep everyone moving, but in slightly different circles, and through slightly different spaces of the spa – minimizing the chance that guests should ever be reminded of the degree to which their well-being might be related to the different life circumstances of those serving them. In this sense, the regimes of mobility set in motion at spas assure the fact that staff members are both tolerated and accepted by guests, but not

10 A 24-hour stay at a Swedish spa begins at approximately 200 US dollars while a basic week-long stay starts at 1,100 dollars. Guests purchasing extra treatments and products quickly find themselves paying significantly larger sums. By and large, spas attract middle and upper class patrons. *Varberg* is something of an exception, since it does still have a medical function, and includes categories of patients whose costs can be covered to some extent by governmental or other institutions.

necessarily *recognized* by them (cf. Dikeç, 2002: 236). The invisibility of others' life circumstances is, it would seem, something of a precondition for the maintenance of a façade of conditional hospitality at Swedish spas.

This is nothing new. Invisibility has long been presumed to be an 'ideal' quality for servants to possess. But as debates about maids and household cleaning services continue to rage in Sweden, it is interesting to note that the world of spas has been spared any similar critical attention. In fact, to the contrary, the spa phenomenon has developed into one of the hottest trends in the Swedish hospitality industry. And eight years after its publication, the quote from the book, *Spa*, which I opened this text with, seems out of synch with the spirit of the times in Sweden today. Where Swedes once dubiously questioned the appropriateness of spoiling themselves without a thought of reciprocating the favour, they increasingly turn to spas for weekend getaways, conferences and longer holidays. In so doing, they turn to a space designed to 'feel' as though it is a space 'for them' and entirely focused upon them. The space of hospitality constructed here is not a space of social relations, but one of inwardly focused consumption, in which Others fade from view. Consequently, it is perhaps not surprising that where Swedes once felt uneasy about paying another person to massage their feet, an increasing number of them have come to accept the idea – without much debate. As it turns out, invisibility through mobility is a powerful tool.

Anxiety and Hospitality

However, trends do not develop in cultural vacuums, and here it should be noted that the rise of spas' popularity in Sweden over the past decade also curiously coincides with an anxiety ridden discourse found throughout the country addressing issues of burn-out, stress and sick leave. It is a discourse in which politicians, doctors and other experts have continuously painted a picture of a public health problem of epidemic proportions, which newspaper headlines would have readers believe is spiralling out of control.

At the same time, the effects of neoliberalism have increasingly made themselves felt in Sweden, leading to a condition in which the welfare state has been put on the retreat. Where citizens once expected to receive support from larger collectivities they are now increasingly left with the responsibility of taking care of themselves. Within this context, it is perhaps not surprising to find that peoples' anxieties have led to the development of 'new' arenas in which they hope to find a hospitable space capable of providing them with a temporary haven from stress, and potential source of rejuvenation. The commercialized hospitality of the spa may be a far cry from any form of pure hospitality, but for Swedish patrons it seems to be developing into a surrogate sanctuary in which they increasingly place their hopes.

It would be easy to dismiss sites of market-driven hospitality, such as spas, as little more than exploitative spheres of labour production and flexible accumulation. Clearly, they are entangled in problems related to these issues. But there is more at issue here that compels us to move beyond this insight and interrogate the question of how these processes are capable of flourishing in contexts troped as hospitable.

A more stringent focus upon the cultural orchestration of mobility can, as I have endeavoured to illustrate above, prove to be an effective means of illuminating the ambivalent and even contradictory processes at work in the name of hospitality. As it turns out, we live in a world in which few doors are opened unconditionally to us. The question is, how are spaces of (conditional) hospitality regulated, monitored and controlled as people enter, depart and move about within them. And what are the processes that are implicated here doing with us, and the manner in which we perceive one another, as well as the world around us?

References

Bacon, W. (1997), 'The Rise of the German and the Demise of the English Spa Industry: A Critical Analysis of Business Success and Failure', *Leisure Studies*, **16**(3), 173–187.

Bauman, Z. (2000), *Liquid Modernity* (Oxford: Blackwell Publishers Books).

Bogard, W. (2000), 'Smoothing Machines and the Constitution of Society', *Cultural Studies*, **14**(2), 269–294.

Brotherton, B. (1999), 'Towards a Definitive View of the Nature of Hospitality and Hospitality Management', *International Journal of Contemporary Hospitality Management*, **11**(4), 165–173.

Deleuze, G. and Guattari, F. (1987), *A Thousand Plateaus: Capitalism and Schizophrenia* (Minneapolis: University of Minnesota Press).

Derrida, J. (2000), *Of Hospitality* (Stanford, CA: Stanford University Press Books).

Dikeç, M. (2002), 'Pera Peras Poros: Longing for Spaces of Hospitality', *Theory, Culture & Society*, **19**(1–2), 227–247.

Falkheimer, J. and Jansson, A., eds (2006), *Geographies of Communication: The Spatial Turn in Media Studies* (Göteborg: Nordicom).

Feld, S. (2005), 'Places Sensed, Senses Placed: Toward a Sensuous Epistemology of Environments' in Howe (eds).

Hallået, R. (1899), 'Hallå! Hallå!', *Ramlösa Hallået* **1899**, 5.

—— (1912), 'Program för pingstfestligheterna i Ramlösa Brunn', *Ramlösa Hallået*, **1912**, 7.

Hasseludden, Y. [website], www.yasuragi.se/, (accessed: 24.09.06).

Heal, F. (1984), 'The Idea of Hospitality in Early Modern England' in *Past and Present* **102**(1), 66–93.

—— (1990), *Hospitality in Early Modern England* (Oxford: Clarendon Press Books).

Hotellrevyn, S. (2003), 'Arbetsvilkor, problem och krav på 5 av landets spa-ställen', 6–7, 22–23.

Howe, D. (2005), 'HYPERESTHESIA, or the Sensual Logic of Late Capitalism' in Howe (eds).

——, ed. (2005), *Empire of the Senses: The Sensual Cultural Reader* (Oxford: Berg Publishers).

Kellner, C. (2003), 'Friskvård gör ansällda sjuka', *Svensk Hotellrevyn*, **6–7**, 22.

King, C. (1995), 'What Is Hospitality?', *International Journal of Hospitality Management*, **14**(3–4), 219–234.

Kurort Hotell and Spa Brochure (2004), Varberg: Varbergs Kurort Hotell & Spa.

Lakoff, G. and Johnson, M. (1999), *Philosophy in the Flesh: The Embodied Mind and its Challenge to Western Thought* (New York: Basic Books Books).

Lashley, C. and Morrison, A., eds (2000), *Search of Hospitality: Theoretical Perspectives and Debates* (Oxford: Butterworth-Heinemann Books).

Lindblad, A. (1907), *Ramlösa* Brunn 1707–1907 (Stockholm: Hasse W. Tullbergs Boktryckeri).

Löfgren, O. and Willim, R., eds (2005), *Magic, Culture, and the New Economy* (Oxford: Berg Publishers).

Mackaman, D.P. (1998), *Leisure Settings: Bourgeois Culture, Medicine, and the Spa in Modern France* (Chicago, IL: University of Chicago).

Mansén, E. (2001), *Ett paradis på Jorden* (Stockholm: Atlantis).

Marcus, G. (1998), *Ethnography through Thick and Thin* (Princeton, NJ: Princeton University Press Books).

Miller, D. (1998), 'Why Some Things Matter' in Miller ed.

——, ed. (1998), *Material Cultures: Why Some Things Matter* (Chicago: Chicago University Press).

Mullins, L.J. (2001), *Hospitality Management and Organisational Behaviour* (Harlow, Essex: Pearson Educational Limited).

Nørretranders, T. (1998), *The User Illusion: Cutting Consciousness Down to Size* (New York: Penguin Books).

O'Dell, T. (2004), 'Cultural Kinesthesis', *Ethnologia Scandinavica*, **34**, 108–129.

—— (2005), 'Meditation, Magic, and Spiritual Regeneration: Spas and the Mass Production of Serenity' in Löfgren and Willim (eds).

—— (2006), 'Magic, Health and the Mediation of the Body's Geography' in Falkheimer and Jansson (eds).

O'Dell, T. and Billing, P. (2005), *Experiencescapes: Tourism, Culture and Economy* (Frederiksberg, Copenhagen: Copenhagen Business School Press).

Pine, J. and Gilmore, J. (1999), *The Experience Economy: Work is Theatre and Every Business a Stage* (Boston, MA: Harvard Business School Publishing).

Reed-Danahay, D. (1997), *Auto/Ethnography: Rewriting the Self and the Social* (Oxford: Berg Publishers).

Selwyn, T. (2000), 'An Anthropology of Hospitality' in Lashley and Morrison (eds).

Sennett, R. (1998), *The Corrosion of Character: The Personal Consequences of Work in the New Capitalism* (New York: W. W. Norton and Company).

Sinnesro (2002), Hasseludden Brochure (Saltsjö-Boo: Hasseludden Konferens & Yasuragi).

Strang, V. (2004), *The Meaning of Water* (Oxford: Berg Publishers).

Swanberg, L.K. (1998), *Spa: Behandling träning mat avspänning* (Stockholm: Raster Förlag).

Telfer, E. (2000), 'The Philosophy of Hospitableness' in Lashley and Morrison eds).

Thrift, N. (2000), 'Still Life in Nearly Present Time; The Object of Nature', *Body and Society* **6**(3–4), 34–57.

——(2004), 'Intensities of Feeling: Towards a Spatial Politics of Affect', *Geografiska Annaler*, **86**(1), 57–78.

Varbergs Kurort Hotell and Spa Brochure (2004), Varberg: Varbergs Kurort Hotell & Spa.

Walton, J. (2000), 'The Hospitality Trades: A Social History' in Lashley and Morrison (eds).

Weisz, G. (2001), 'Spas, Mineral Water, and Hydrological Science in Twentieth-Century France', *Isis*, **92**(3), 451–483.

Wood, R. (1994), 'Hotel Culture and Social Control', *Annals of Tourism Research*, **21**(1), 65–80.

Home (Hackney, London, UK, 2006) by Elly Clarke

Chapter 7

Resident Hosts and Mobile Strangers: Temporary Exchanges within the Topography of the Commercial Home

Paul Lynch, Maria Laura Di Domenico and Majella Sweeney

The commercial home, where 'paid for' accommodation is provided in a property that also functions as a private home, is an enduring, popular and complex social phenomenon. Indeed, its various guises and manifestations such as the bed and breakfast (B&B) or guest house provide an enduring and perhaps endearing image of traditional British hospitality. Against this backdrop and setting, this chapter explores the nature of the commercial home setting and the motivations of both host and tourist. We look at the reasons that compel each actor to purposefully seek out and interact with the other. The proprietor accommodates the traveller who in turn chooses to stay in the home of another rather than other more impersonal sites of tourist accommodation. The encounter is part of an important journey not only for the tourist on the move but also metaphorically for the resident host. By revealing the motivations and drives of each in relation to the tourist's visit to the host's home, we unearth a dynamic interaction which is imbued with notions of both the permanent and the ephemeral. The steadfast solidity of the home and its residents with their local roots and sense of belonging are overlapped by the transitory interludes of the temporary guest. The transient presence of the stranger injects dynamism into the home and the world of its more permanent residents, shaping the memories and fluid processes of the transaction found therein.

This chapter is in many ways a textual tour of the commercial home, a setting in which the host and guest interact with each other.[1] The commercial home is

1 This chapter emerges from our reflections upon the commercial home experience through a 'mobilities' lens and is empirically illustrated by our respective research into the commercial home host (Di Domenico, 2003) and guest (Lynch, 2003). We also draw upon recent empirical insights and collated images contributed by Sweeney (ongoing research) on the host's conception and attachment to place. Where quotations from proprietors are cited these derive from depth qualitative interviews carried out with hosts in their own home (Di Domenico, 2003). Where guest-orientated accounts are provided these are derived from guest-researcher observations of the commercial home experience (Lynch, 2003). This sociological impressionist approach (Lynch, 2005a) may be likened to an exploration of both the 'inner' and 'outer' journeys of the individual (Myerhoff, 1993) with the researcher being concerned with exploring and analysing their 'own' experiences, interactions and associated observations. This method fits with the type of mobility research argued for by Sheller and

Table 7.1 Broad Characteristics of Commercial Home Enterprises

	Smallest Commercial Home Unit e.g. Host Family	Medium-sized Commercial Home Unit e.g. Guest House	Largest Commercial Home Unit e.g. Hotel
Number of Bedrooms	1-2 Bedrooms--- ----------------------------- →15 Bedrooms for Guests for Guests		
Room Occupancy Levels	Lowest --------------------------------------- →Highest		
Host perception of Commercial Home Enterprise	Private ------- →Commercial Home ------- →Business Home		
Entrepreneurial Orientation of Host	Least --- →Most Entrepreneurial Entrepreneurial		
Economic Dependency on Hosting Income	Low Economic ------------------------------ →High Economic Dependency Dependency		
Most Common Primary Host Gender	Female-------------→Female/Male--------- →Male/Female		
Degree of Partner Involvement	Least Partner -------------------------------- →Most Partner Involvement Involvement		
Family Participation Level	Family --------------------------------------- → Family-run Involvement		
Higher Education Experience frequency	Least Frequency with ----------------------- → Most with Higher Education Higher Education		
Host Training Orientation	Lowest --------------------------------------- → Highest		
Host Engagement with Home	Greatest -------------------------------------- →Least Engagement Engagement		
Home-Host Relationship	Home as Reflection ------------------------ →Home as Reflection Of Self Highest Of Self Lowest		
Product Commodification	Least --- →Greatest		

Source: Lynch, 2005b

emblematic of the interplay between dwelling and mobility discussed by, for example Clifford (1997) and Urry (2000); Urry (2000: 157) refers to people being able to 'dwell in various mobilities'. Its counterpoint from a commercial home host perspective explored here is that of 'mobility-in-dwelling' where the nature of activities in the dwelling, facilitated by the relationship between humans and inhuman objects, enables various forms of mobility. We will take you, the reader, on this journey into the home of the host. As temporary guest, and transient interloper, we will guide you through this intimate yet commercially-orientated dwelling where you will experience many facets of home that will seem both familiar and strange to you. The areas we discuss may be seen as a deconstruction of the complex interplay between dwelling and mobility. First we begin by unpacking the nature of the setting as bound by the domestic context. We then punctuate the journey with insights into the standpoints and motivations of the resident host as well as other guests who similarly choose to venture into the home of another. The third part of the chapter provides a tour of the home and insights into the hospitality transaction by looking at virtual representations prior to the visit and experiences during the stay itself. In the fourth section we elaborate upon the importance of artefacts within the commercial home and their importance as signifiers of both identity and place. The fifth part of the chapter elucidates the theme of social interaction and control. Here we consider the inherent inequalities in the relationship between host and guest. We conclude by drawing together our insights of this particular manifestation of commercial hospitality, its inherent tie to the home domain, and the paradox of enduring mobilities to be found therein.

The forthcoming tour may prompt you to recall some of your own actual previous journeys of staying in such privately-run commercial homes. Perhaps some of you may even have adopted the role of host, and so can recall awaiting the arrival of the guest. We hope this tour does indeed stir such memories or makes you recall similar former experiences, and we urge you to revisit them alongside these new ones as we take you through the topography of the commercial home.

The Commercial Home

We use the term 'commercial home' in order to embrace a range of accommodation types including some (small) hotels, bed and breakfasts (B&Bs), and host family accommodation which simultaneously span private, commercial and social settings (Lynch, 2005b). Thus, it refers to types of accommodation where visitors or guests pay to stay in private homes, where interaction takes place with a host and/or family

---

Urry (2006) that is concerned with how people (that is the researcher) interact with place (that is the commercial home), events (that is the experiences) and people (that is the host(s) and host family). It captures the complex interplay and interrelationships formed between people and place, drawing out the significance of not just conversational interactions but also aesthetic, corporeal and sensory transactions. Similarly, Di Domenico's (2003) conversational interview approach with hosts allowed the researcher to explore host-centered accounts and self-presentations enabling a penetration of forms of expression and identity using the actual words and accounts of the hosts themselves.

who usually live on the premises and with whom public space is, to a degree, shared. Our aim in this regard is to use an inclusive term which places emphasis upon the use of a functioning private home and permanent residence for long-standing occupants. For all visitors it is a sharing of space, often involving close interactions between the host/family and interim guests. The concept of 'home' is therefore dominant, imbuing the varieties of interactions that take place. The duality of purpose of the setting as both a private residence for its owners and a venue for paid-for accommodation for visitors is thus evident.

Table 7.1 provides an overview of broad characteristics of the commercial home against which the reader may wish to locate aspects of the discussion presented in this chapter. This summary is useful when considering the range of possible permutations of business-home accommodation hybrids that exist, their topographical features and characteristics pertinent to the host and/or hosting family.

The fusion of intangible and tangible elements should be noted. The commercial home is an example of a hybrid system that recognizes the 'complex relationality of places and persons connected through performances' (Sheller and Urry, 2006: 214) and the combination of 'objects, technicalities and socialities' leading to the production and reproduction of distinct places (Ibid: 214). The commercial home is an enduring global social phenomenon that dates back to the earliest days of hospitality. Typically, commercial homes will have between one and 15 rooms used for letting purposes. Even in countries where hotel chains have significant market penetration, such as the United Kingdom, commercial homes are numerically the single most prevalent (but frequently overlooked) form of commercial accommodation. Studies have identified the positive local, economic and social contributions of such small enterprises as compared with larger hospitality firms (Kontogeorgopoulos, 1998; Andriotis, 2002).

Commercial homes may be contrasted with the so-called *non-places* of modernity (Auge, 1995) such as motels, the latter very often designed to a standardized formula in order to befit the needs, expectations and likely stop-off locations of the commercial traveller. Commercial homes may be deemed to represent the quintessential *place* of modernity, whether urban or rural, local or cosmopolitan, in that they are frequently perceived to embody a kind of authenticity (Ritzer, 1996), where the concept of the 'local' is partly metaphorical, communicating the idea of bounded local communities with 'intense social relationships… characterised by belonging and warmth' (Urry, 2000: 48). Thus, they sit in counterpoint to the McDonaldization of hospitality, are symbolically anti-hotel/motel and their authenticity together with their social embeddedness in local communities (Ateljevic and Doorne, 2000), leads to their being associated with the reinvention of a kind of local identity, which contrasts with late modern permutations of globalization. Through our situated tour of interactions within the space of the commercial home, we wish to critique dominant notions of mobility and the specific articulation of 'mobilizing hospitality', as the overall theme of this book. The commercial home as a specific and physically immobile dwelling is *de jure* representative of more sedentary conceptualizations of place and movement (Sheller and Urry, 2006: 214). However, our interest lies in *de facto* mobility as manifest by social interactions. Therefore, it is not only the traveller who is the mobile and iterant protagonist. We argue that both the host and the guest embark upon a journey, and are engaged in social activities framed by the act of proffering or consuming hospitality as determined by their respective standpoints. Fixed spaces therefore can act as canvases for new social encounters and engagements with the 'other', facilitated by the opportunities afforded by new technologies and improved transportation links. In this chapter we not only highlight the intrinsic complexities of offering hospitality on a commercial basis within a home setting, but demonstrate that this is also a dynamic fixed site of inherent mobility.

These graded (by national tourist boards and marketing organizations) and often un-graded forms of accommodation are performative settings, territorial nodes, a second line of visitor accommodation (in the United Kingdom), providing less controlled and controllable (by tourist boards seeking to homogenize quality assurance standards) visitor experiences than 'first line' hotels and motels. They are in some ways similar to but also different from hotels, airports and other venues frequented by the traveller. They give rise to a distinct social space, constructing contemporary complex forms of social life in a world where time is compressed and the nature of human relationships has been transformed, especially for the more affluent members of society. The purchase of the commercial home is frequently in itself an act and symbol of upward mobility where commercial letting makes affordable a home property and enables access to a location and potentially a lifestyle that might otherwise be unachievable. Two broad types of purchase may be distinguished, one where a property is ostensibly a private home then used additionally for commercial letting purposes, another where a property is purchased owing to its previous conversion for commercial letting purposes.

Figure 7.1 deconstructs the various elements of the commercial home product. The framework acknowledges the emotional significance of the setting and recognizes its role as a performative player. It highlights the process of the construction of the commercial home product by guest and host.

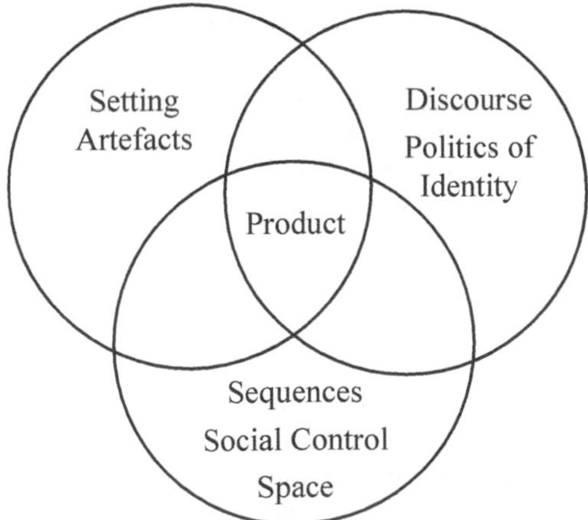

Figure 7.1 Construction of a commercial home product
Source: Lynch, 2005a

The Commercial Home Space

The definition of the commercial home refers to the sharing of (public) space. In this sense, a dramaturgical use of the public/private spatial categorization is employed in order to distinguish space. The space that is used to some degree by hosts and guests can be viewed as public space in opposition to space used exclusively by the long-term occupants of the property, for example, a bedroom, bathroom or kitchen, or on a short-term basis by the guests, such as, the bedroom. However, spatial usage within the commercial home is multifaceted and might be better viewed in line with Sheller and Urry's (2003: 108) description of 'a more complex de-territorialization of publics and privates, each constantly shifting and being performed in rapid flashes', that is a spatial mobility.

The size of the property, in conjunction with the frequency and volume of hosting, determines the degree of spatial commodification and consequent clarity of the division between public and private space. Thus, in a small family-run hotel with up to 12 letting bedrooms, the physical division is likely to be very clear whereas in a cultural homestay, one might find no signalled distinctions between host and guest space. In the absence of signs distinguishing space, more subtle spatial markers may be found, for example, a door half closed may communicate usage by the host requiring knocking on the door by the guest to obtain permission to enter (a house rule typically discovered by trial and error), or, as is often the case, room usage may be temporal with 'ownership' shifting between hosts and guests several times during the course of a day. Guests seeking to make use of a sitting room may find, for example, the host's children watching a television programme as they re-territorialize 'guest' space in line with their usual daily routines. In so doing, their private behaviours are made public to the guest who is then presented with the dilemma of retreating to their bedroom, enforcing guest usage or negotiating a shared usage. The nature of artefacts and décor frequently communicates spatial 'ownership', with objects more personal to the host found most frequently in space they habitually use, and conversely, spatial depersonalization in space allocated more exclusively to guests. Language is also an indicator and gives rise to some ambiguity as noted by Lowe's (1988) observations regarding host usage of the term 'dining-room' instead of 'restaurant' in a small family hotel.

Not only does the commercial home challenge traditional conceptions of public/private space owing to its contested and fluid usage, but it also does so for the home/hotel dichotomy (Douglas, 1991) since the commercial home provides a bridge between pure forms of the private home and hotel. Similarly, Treadwell (2005) describes the motel in relation to its inherent anonymity, its transit form and low level of homeliness drawing out its opposition to the concept of the home and yet, as she argues, despite this, the motel provides elements of the home, for example, traces of past occupants, such as the stray hair. A further similarity to the home is Treadwell's description of the motel as a space where 'the fixed and the mobile coincide' although such a semblance is 'treacherous' (2005: 216). While such contrasts have utility, and indeed we make use of them here on occasion to highlight aspects of our argument, the commercial home nevertheless challenges their uncritical acceptance.

Firmly Rooted and on the Move: Unpacking Motivations of Host and Guest

Aramberri (2001) is critical of the host-guest paradigm with its consequent associations of host/guest protection, reciprocity/exchange and duties for both sides, owing to the commercial nature of modern mass tourism rather than being an essentially non-monetary exchange. Recognition is given to the strengths of this argument, and, as discussed later, there is sympathy with the unsatisfactory nature of the exchange concept. However, the authors consider the peculiarities of the commercial home with its straddling of both private and commercial domains, its enactment of behaviours that conform less to 'providers of services' and 'consumers' in modern mass tourism (Aramberri, 2001) and that comply more to the concepts of hosts and guests in modern independent tourism in which many cultural rituals of traditional domestic hospitality are enacted. Therefore, the host-guest terminology is employed.

The Commercial Home Host

As Urry (2000: 132) suggests, 'contemporary forms of dwelling almost always involve diverse forms of mobility'. Like the motel (Treadwell, 2005), the commercial home is connected to broader scapes supporting travel and admits flows of people for normally short-lived periods, and often on an intermittent basis throughout the year. Anchored to the performative stage of the commercial home, the host is able to perform a multiplicity of 'mobile' roles. For instance, hosts may embody many facets of the local community and its traditions; they may be an important 'hospitable ambassador' to visitors from abroad in shaping national and local representations of place; and they often perform the husband/wife, parent and/or neighbour roles which support and uphold the values of the community within which they reside and which they may be seen to represent (Di Domenico and Lynch, 2007). The complexity of the concept of community in the contemporary world (Urry, 2000) is reflected in the various ways hosts represent their relationship, a part of their politics of identity, as well as reflected in relation to objects and associated discourses. Hosts may themselves be strangers to their 'local' or 'national' community but nevertheless interpret their host role as ambassadors and interpreters based variously on 'propinquity', that is topographical, 'communion', that is characterized by close personal ties or 'localness', that is bounded social system (Bell and Newby, 1976) and only very rarely all three senses of community in combination. For example, the views of the following host show that, despite being a relative newcomer to the area in which her guest house was located, she was intent upon establishing close links with the local community and others with similar interests to herself:

> I went round to all my neighbours, to their door. I walked in and introduced myself. Everyone is really friendly… I also became very friendly with a lot of the landladies along the road and this [the guest house] is one of the largest ones… we all help each other out. If I'm full then I'll phone one of them up and if somebody else is then they'll do that too. So it works quite well…Yeah. Friendly. Definitely. Two of the couples were at my daughter's wedding. In the evening we go out. We don't meet up much for coffee during

> the day because we're all so busy… Since I came on the scene I organised our Christmas night out for eighteen of us. So we'll probably do that this year again.
>
> (Female host – Di Domenico, 2003)

The community relationship is also often communicated to guests via discourses concerning objects, for example, a picture purchased from a local artist, or via discourses concerning the age of the property, its previous occupants, and possibly reinforced by a preservation order or its location in a conservation zone. Such behaviours are open to various interpretations but the point we would make here is the often somewhat ersatz nature of the invocation and communication of community in response to the receipt of paying guests. In other words, a rather mobile host-community relationship. Such multiple roles stress different dimensions of their politics of identity and respond to as well as shape audience expectations. In the case of the commercial home, the significant additional factor is that the home is used for commercial hosting. The admission of people into the otherwise private home requires adaptation of space, its control, presentation and usage often for brief periods of time whilst simultaneously everyday (non-commercial hosting life) carries on.

The host uses their home as a mechanism through which the individual is able to 'effect agency' (Urry, 2000: 78) and accomplish a range of mobile needs: physical, social, economic, psychological, temporal, communicative (Sweeney and Lynch, forthcoming). The overarching accomplishment that the home enables is the delivery of a certain 'lifestyle'. The home is not simply a physical shelter. It embodies the dreams, hopes and aspirations of the host. It is a symbol of status within the local community, a personal and social anchorage, a place of meetings, of comings and goings, an immobile yet malleable stage for performances and interactions. It may serve as a family home, as a repository of memories, as a link back to the past. In its presentation it communicates information on welcome, on the type of stay that is being offered. Thus, as the labelling chosen communicates the genre of property used, such as those of 'townhouse', 'country house' and so on, this also sends important messages to the potential visitor and acts as a mirror of the tastes and personalities of the inhabitants (Marcus, 1995). As a commercial home, it is an economic tool for the present but also serves as an asset providing the promise of financial security in the future.

In small commercial homes the proprietors are frequently female. As the enterprise becomes larger, there is often a greater engagement by male hosts. Thus, on the one hand, commercial home hosting both historically (Walton, 1978) and on a contemporary basis (Di Domenico and Lynch, 2007) may represent a means of gendered social and economic emancipation. On the other hand, the distribution of tasks amongst hosts may simply reinforce gendered roles. Therefore, the commercial home site represents both social mobility *and* immobility, akin to the concept of mobile immobility of Kesselring (2006: 275) and provides an example of mobility practice 'structured by contextual situations, economic and social conditions, and power relations in general' (Ibid: 270). Hosting international visitors may be motivated by a desire to transcend actual physical immobility in order to attain heightened social status on a neighbourhood basis as seen in the example of families of modest incomes that host international students who study English as a foreign language (Lynch, 1998). In some communities, the hosting of international visitors is identified as bringing added status as the host is seen to possess international networks (Cole, 2007). In addition, hosts with the densest social networks are identified as being looked up to as role models for their ability to maximize volume of visitors and transform 'pin money' into a significant income stream permitting a higher standard of living. Thus, the income may be used, for example, to fund their children through private education, to book additional international or long-haul holidays for themselves and their families, or to purchase an extra car. Most important of all, the hosts are able to possess a home of a more prestigious standard than they would otherwise have enjoyed. The recognition of the importance of the commercial home in terms of the lifestyle it affords to its occupants was repeatedly acknowledged by hosts, as demonstrated by the following;

> We could only really afford this large house if we ran a B&B. I'd always dreamed of living in a place like this. Well, there'd be no need for all the extra rooms and we'd be rattling about the place. I love the large house, the garden, and the area is great for bringing up kids. But it wouldn't have made sense and wouldn't have been possible if we didn't have the business. They go hand in hand really.
>
> (Female host – Di Domenico, 2003)

For the guest, the host is a stranger whom they encounter and seek out in their quest for temporary accommodation and sustenance. However, the host is also the insider within the commercial home where the guest is the stranger who in time too may become a temporary insider as allowed for by the social conventions and laws of hospitality. On occasion a deeper relationship may evolve, one of friendship surviving beyond the closure of the hospitality transaction. Social engagement with guests, the receipt of gifts, the receipt of stories of their home country, and post-stay correspondence all permit the host to indulge in vicarious mobility that transcends the reality of physical immobility, and on occasions is a subsequent vehicle to actual mobility where invitations to visit the guest's home are accepted. This perspective is reflected in the views of a host who with his wife ran a small guesthouse from their home in the suburbs of Inverness:

> This is not just a business to us. A lot of the people we've had to stay have become friends over the years. Okay some people you never see again but others you do strike up a friendship with if they're here long enough and you get on. You can see that by the Christmas and Easter cards we get every year and the invitations we've had to family get-togethers... We actually went over to Spain to visit a student who'd been with us for about three months over a year ago. Well, we were actually on holiday there and popped in to see her and ended up staying an extra 3 days.
>
> (Male and female hosts – Di Domenico, 2003)

Thus, the commercial home is both home *and* away owing to the effects of hosting. It implies simultaneously both notions of the sedentary *and* the mobile.

The Commercial Home Guest

The limited evidence available regarding commercial home guests suggests that in economically developed societies they increasingly tend to be more affluent (Tucker and Keen, 2005) and reflect a tourist profile not infrequently international (Ferguson and Gregory, 1999) that may have particular interest in or display a greater concern with the rights and ethics of the local community (Bauman, 1993). Tucker and Keen (2005) identified most guests as between 30 and 69 years, although in the case of host families (Lynch, 1998) the majority of guests are students aged between 18 and 25 years. From observation, commercial home guests are usually couples, single or, less frequently, in family groups. Length of stay varies, typically between one and three nights, although in the case of guests of host families the average length of stay is three to four weeks. There are also more regular 'repeat guests' and long-term resident guests who may stay for extended periods due to work and other commitments based locally. Travellers who select such accommodation may do so for many different reasons. The decision may of course be simply one of location or cost. They may wish to obtain what they consider is a good service for a modest outlay or there may in fact be little choice of alternative accommodation providers particularly in more remote destinations. However, the visitor may also be more discerning in their choice of where to stay and may seek out what they regard as a more 'homely' or 'genuine' form of hospitality as an antithesis to, or even respite from, more standardized types of branded hotel chains. Motivation may stem from a desire to ensure that money goes to the local community rather than to the corporations that may be located thousands of miles from the accommodation provider. Such tourists deliberately seek out the non-McDonaldized accommodation experience (Ritzer, 1996) where the idiosyncratic nature of the accommodation experience, and the potential for seemingly authentic interactions with representatives of the local community, are important parts of decision-making criteria. By staying in such accommodation, the guest may be symbolically rejecting corporate accommodation experiences. Such guests fit closely with those characterized as 'global cosmopolitans' who maintain 'an ideology of openness towards certain "other" cultures, peoples and environments, often resulting from extensive corporeal travel' (Urry, 2000: 173). This engagement with 'other' cultures is complex in that the 'other' within and between economically developed societies will lead to stays in homes whose inhabitants' lives are often not so very qualitatively different from their own. Thus, this type of guest may be both physically nomadic and experientially sedentary.

The guest is the stranger, the 'other', whose presence in the commercial home, however intermittent, gives rise to effects far beyond simply that of engaging in economic transactions. The home is for them a destination, a retreat, but also a place of transition, of comings and goings, a place where they will be managed by its normal occupants through their rituals, routines and norms. Conventions regarding the timing of meals, the normal bedtimes of the host family, and social codes pertaining to noise, decorum and other protocols mould the experiences of the guest and their sense of ease. Thus, the effects upon guests may be positive as well as negative, whereby guests must negotiate their way through the unknown rules of social conduct prescribed by the script of consuming hospitality within a domestic setting.

A successful experience occurs where the guest's subjective tastes, expectations, cultural and behavioural norms coincide with those of the host. Simple, even mundane, yet emotively powerful examples are the guest's reactions to the food made available to them, the décor they encounter, or positive responses to the facets of the politics of identity presented by the host. Thus, it is often the case that 'harmonious and disharmonious' hospitality realities exist side by side. For example, the following excerpt is from the guest-researcher's tape-recorded impressions in relation to a guest house room:

> The room itself is the smallest that we have been in so far… it has got a single bed and a double bed, and each bed [has] a half moon Formica shelf, there are two wall-mounted lamp fittings, sort of hotel type… we had to relocate an armchair in order to set up the cot… there is a dressing table, the noise in the background is a power saw… and there is a wardrobe… it's got a radio clock on top, which looks like it must be out of commission, the Formica surrounds underneath the bay window, there is some damage visible to the fabric in terms of the wallpaper above the window and on the ceiling… I have worked out the damaged corner of the shower is from tall people trying to leave the shower banging their head against it. It's also got a luggage rack, curtains of velours textile, curtain rail, badly fitting one has to say…
>
> (Excerpt from guest-researcher's diary – Lynch, 2003)

Although the accommodation fell below the guest-researcher's expectations, the quality of the interactions with the hostess and her natural hospitableness were appreciated:

> Mrs W. is about 60 odd, the description pleasant, matronly, person comes to mind… here we have real conversation, genuine interaction taking place…We had celery soup, followed by roast chicken and wonderful roast potatoes, which we complimented Mrs W. on, and found out that she boils them for ¾ [of an hour] and puts them into very hot fat to roast them… she was concerned we weren't too cold, she found the heating difficult to adjust with the weather, and she was solicitous of our welfare…
>
> (Excerpt from guest-researcher's diary – Lynch, 2003)

Touring the Home: Host-guest Permutations and the Hospitality Transaction

The nature of human interactions in the context of hospitality is usually conceptualized as an exchange. For example, Lashley (2000) draws upon the ideal of reciprocity that inherently underpins the hospitality experience (Telfer, 2000). However, the interactional reality of hospitality production and consumption is such that it is better conceived as a multi-faceted and quintessentially 'mobile' transaction, encapsulating facets of an economic, social, cultural and psychological exchange between actors. 'Transaction' conveys the idea of a 'crossing over' between host, guest and place (Lashley, Lynch and Morrison, 2007).

Before the Arrival: Shaping Hospitality Illusions and Expectations Through a Virtual Lens

The Internet has become an empowering tool both for travellers and for many small hospitality businesses (see also Germann Molz, Chapter 4 of this volume, on mobile hospitality and the Internet). It has enabled the microenterprise to transcend its resource insufficiency to promote itself on a global basis. Internet promotion is especially relevant to the commercial home as the medium enables virtual presentation of the host and family, as well as permitting a virtual tour of the accommodation itself. The power of the Internet in allowing the proprietor to shape the expectations and experiences of the guest even before their arrival is now made possible. This promotion places even the most remote and small-scale B&B on a par with larger corporate hotels which are constantly at pains to communicate their brand to reach potential markets. Proprietors are acutely aware of this new resource and often describe the advertising facilities that they use not only as a means of increasing awareness and market penetration, but also as a means to disseminate a certain image reflecting their uniqueness:

> I think people choose me because of me because I have my picture on the website and we have a… not a pompous sort of website you know, we have on our website that there's a couple of old codgers who do B&B and who love it and who never want to stop, and, you know people can tell that it's a relaxed kind of place that they're coming to.
>
> (Male host – Sweeney ongoing research)

Commercial homes are often associated with lifestyle businesses (Morrison, 2002; Shaw and Williams, 2004) and communication of the lifestyle image is thereby enabled by the Internet. In exhibiting a property, the host reveals information about themselves and the household's occupants which assists the guest in deciding whether to stay.[2] Tucker and Lynch (2004) describe how Internet promotion may portray a psychographic 'lifestyle' profile of the hosts and the potential nature of the stay. It conveys information about the sociable nature of the hosts and signals high levels of host-guest interaction. It may also indicate the interests of the hosts, such as gardening, photography and local history, and may show household artefacts in conjunction with explanations about their personal and historical significance. The guest uses and distils the information in order to apply an 'imaginative mobility' (Haldrup and Larsen, 2006: 281) in which they conceive of themselves as a guest engaging affectively not only with the property but also with the hosts and their lifestyle. This may be likened to a form of host-guest Internet dating site where users select their potential host date. There is resonance here with Bell's (2004) observations concerning how people may walk into a restaurant and recognize that this is 'our kind of place' (46) owing to taste, for example, of property, artefacts, self-presentation, being used as a marker of recognition.

---

2 For example, one Scottish tourism award winning website promoting guest houses includes photos of the hosts, their pets and previous guests. See http://www.gems.scot.info/specgraphics/special.htm.

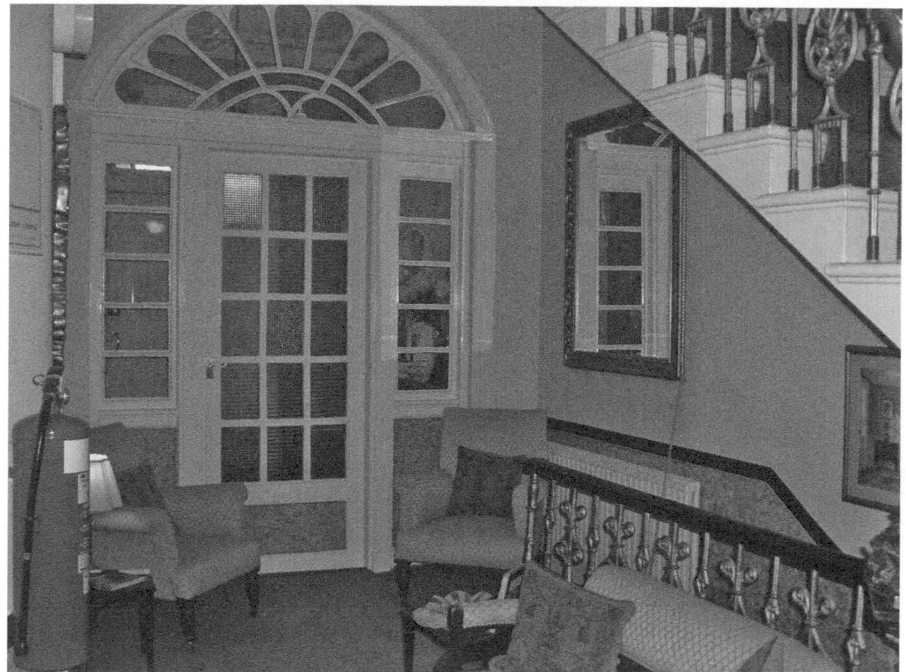

Photograph 7.1 A sense of homeliness or institutionalism?
Source: Majella Sweeney

Across the Threshold: Experiencing the 'Visit'

In order to explore the dynamics of the hospitality transaction within the particular setting of the commercial home, we will elaborate further on each element of the commercial home product identified in Figure 7.1. Guest impressions of the commercial home building, its nomenclature, and the locality are all important in stimulating the feelings and imagination of the observer. The setting has a behavioural effect provoking social conformity, fostering role-plays, leading to rule interpretation, but may also create dissonance. Artefacts similarly convey product information which intersects with a live domestic setting and family home. The domestic setting and its associated symbols thus engender interest in the householders and also communicate information on their lifestyle and politics of identity. As the following illustration depicts, artefacts determine the product and communicate, for example, a sense of homeliness or institutionalism, or even bemusement as to what is on offer.

Artefacts may highlight ambiguities and blur the boundaries further between our notions of domestic and commercial hospitality. We are prompted to question our pre-existing suppositions regarding these two forms. The former is often associated in our minds with unremunerated hospitality punctuated by family, kinship and obligation. The latter we associate more with a formal and paid-for service, for which we voluntarily offer custom and cash. Thus, the familiarity of such displays may yield in us a confused sense of trepidation driven by our inability to reconcile and clearly distinguish between these two forms of hospitality.

External 'imposition' of artefacts by, for example, tourist boards are also found (Photograph 7.2). These engender further boundary confusion by their association for us with the more impersonal and bureaucratic hotel regulated by quality checks and service standards. Here too these traits are evident, providing us with reassurances of some more neutral force ensuring a degree of standardization and levels of hygiene in which we are meant to place our trust. Yet these punctuate our experience with a sense of disquiet at the intrusion of these counterpoints to the domestic.

Discourse themes may be remarked in relation to the setting, artefacts, identity and product, and of three types: 'authentic', social role and sequential (see Figure 7.1). Commercial home participants employ discourse as a mechanism for role casting. Such discourses collectively serve to construct pictures of the hosts, their families, interests and activities, as well as their life events. Authentic discourses are characterized by the converging realities of the participants, where social role-playing is apparently missing. Most discourses occur during structured sequences of activities, such as mealtimes. Politics of identity can be classified as 'ascribed' and 'appraised' to communicate the way in which on the one hand guests (and indeed hosts) may categorize the host and on the other to assess the experience of the quality of interactions. 'Ascribed' refers to characteristics such as gender, social class, and so on. 'Appraised' refers to subjectively appraised characteristics such as welcoming

Photograph 7.2 A domestic hallway transformed into a more commercial reception area.

Source: Majella Sweeney

or not. Appraised characteristics are identified as important contributors to product selection and expectations, as well as the most important of product elements.

Sequences assist in organizing social life. Four types exist: structured host-guest sequences, unstructured host-guest sequences, host family sequences, and guest personal sequences. Structured host-guest sequences, for example, the showing of the bedroom. Unstructured host-guest sequences include host-guest encounters and conversation. Host family sequences may also affect the guest's stay. Guests have their own sequences contributing to the co-construction of the commercial home product.

> A. and I had planned our return around the fact that we would be settling D. [child], which meant that we would time our pre-dinner arrival for the aperitif like last night… and just because it really suited us… but I suppose it does suggest a control strategy on our behalf…
>
> (Excerpt from guest-researcher diary – Lynch, 2003)

A wide array of social control methods are employed by hosts in ordering interactions, including spatial allocation and associated temporal usage, and the temporal organization of meals. Social control may occur through the organization of artefacts, for example, in the presentation of guest areas. Control may occur through the 'presence' of the absent owner, or be imposed externally to the commercial home unit. An array of spatial control strategies exists, the intensity of which varies by type of commercial home, whether large, for example, a small hotel or small, for example, a B&B. The balance between home and enterprise also varies and is evident in fluid permutations.

In combination, the elements outlined in Figure 7.1 impact upon expectations, product nature, tensions, and guest behaviour. Product expectations are stimulated through advertising materials and then further indicators are picked up on through guest interactions with the product and hosts. Therefore, divergence between expectations and experience is quickly noticed. Ideally, there is coherence between all product elements. Its absence may cause uncertainty as to suitable behaviour. The experience of the home product differs and provides interest for both guest and host. Length of stay is important in engendering a sense of home. As we have shown, a number of home product tensions exist. The provision of commercial hospitality in a private home is intrinsically antithetical owing to the domestic/commercial duality. The domestic setting affects behaviour and constrains conduct that would not occur in a purely commercial hospitality setting. Such tensions contribute to the adventure of the experience and form a perverse attraction for the guest. The co-presence of host and guest provides special opportunities for engagement in the transaction between host and setting, recognized as an ideal. Such participation contributes to the guest imagining long-term residence in the home. However, whilst the guest to a degree enters into the daily existence of the host, the presence of the guest disturbs the house's residents leading to new social behaviours in the home.

Home as Origin and Destination: Conversing with Artefacts

In the previous section we interrupted our journey through the commercial home with brief reflections upon its contents and artefacts. We wish to return to such issues at this point. Our interest in artefacts lies in the communicative codes of objects or other symbolic displays within the stage of the commercial home. Such artefacts have a performative role in that they embody the self of the host and the other household occupants whom they mirror, communicating messages about the individuals ready to be deciphered by the guests. In this sense, the home setting and its occupants act as a spectacle even in the absence of the host (Lynch, 2005a). Such objects also enable some engagement or personification of the locality and the nation. This may be marked where hosts attempt to communicate their own sense of identity through their own attachment to place as illustrated by the following views of a guesthouse proprietor from the Highlands of Scotland:

> They want to see Loch Ness or the Castles. It's a very important part of our heritage and I don't think enough is made of that. I think the tourist board recently tried to get away from the heather and tartan image, but you are never going to get away from that because that is what Scotland is all about. That is what people want to see… my guests like the tartan wallpaper. Now some might think it gaudy but I think we should be proud of the Scottish traditions. That is the image which is projected all over the world.
>
> (Female host – Di Domenico, 2003)

Furthermore, objects convey information regarding the commercial home that is part of a broader connected 'scape' (Urry, 2000: 35) of places, peoples and organizations, such as tourist boards, most visibly through the presence of official paraphernalia of recognition (Photograph 7.3). The guest encounters these connections of the outside world when they view certificates and plaques on the walls or well-signposted fire doors. Such externally imported and imposed artefacts contribute to the multiple identities that the commercial home must perform. It is simultaneously a family home, a site of commercial accommodation, an economic asset and a means of future financial security for its owners, and a meeting place or neighbourhood/community dwelling. Indeed, it is important also to appreciate the importance of travel away from the home location for the host as well as for the guest. Hosts may display artefacts which depict scenes of their previous lives and residences elsewhere and betray their own desires to visually recount to the stranger that they too have travellers' stories to tell (Photograph 7.4).

This presentation may create both a sense of fascination and one of ease for the guest and a greater sense of empathy for their host. Alternatively, they may present a more eccentric façade which is not altogether in keeping with their expectations of a more stereotypical view of 'homely' décor. Thus, the commercial home space may act as a dynamic and fluid repository for the storage of objects which allow for the consuming of alternative places far removed from the setting which they occupy.

The way in which domestic space is adorned and presented affects the performance within it. Objects 'tell stories' to the guest, whether in fact real or imagined, and may provide a basis for stimulating conversations with the host, thereby facilitating

Photograph 7.3 Object reflecting the hosts that stimulate guest interest and enable host story-telling
Source: Majella Sweeney

social interaction. On one level, such artefacts can be read as signs (Osborne, 2000), on another level as objects stimulating neuro æsthetic responses. The portrayal of the domestic sphere also elucidates its function as a lived mechanism bound by underlying cultural logics and meanings which derive from the outside world that are in turn interpreted by its inhabitants (Madigan and Munro, 1991). Such symbols or artefacts provide the home with an active voice with which to communicate. Thus, both the guest and host interactively create, instil and engage with meaning from the space of the commercial home. One can contrast here the commercial home with the hotel, the latter of which does not reflect the personality of a specific individual or householder, and is therefore an arguably less engaging sense-scape.

Inclusion or Intrusion? Inequalities of Transient Commercial Hospitality in the Home

In the commercial home it is the host who operates the rights of inclusion and exclusion (Lashley, Lynch and Morrison, 2007), to proceed or not with the 'blind' encounter with the impending, anticipated guest. The ethnocentric host desires social interaction on his/her own terms rather than the uncovering of the guest's true identity. Guests lack power in deciding the social control of initial access as they first

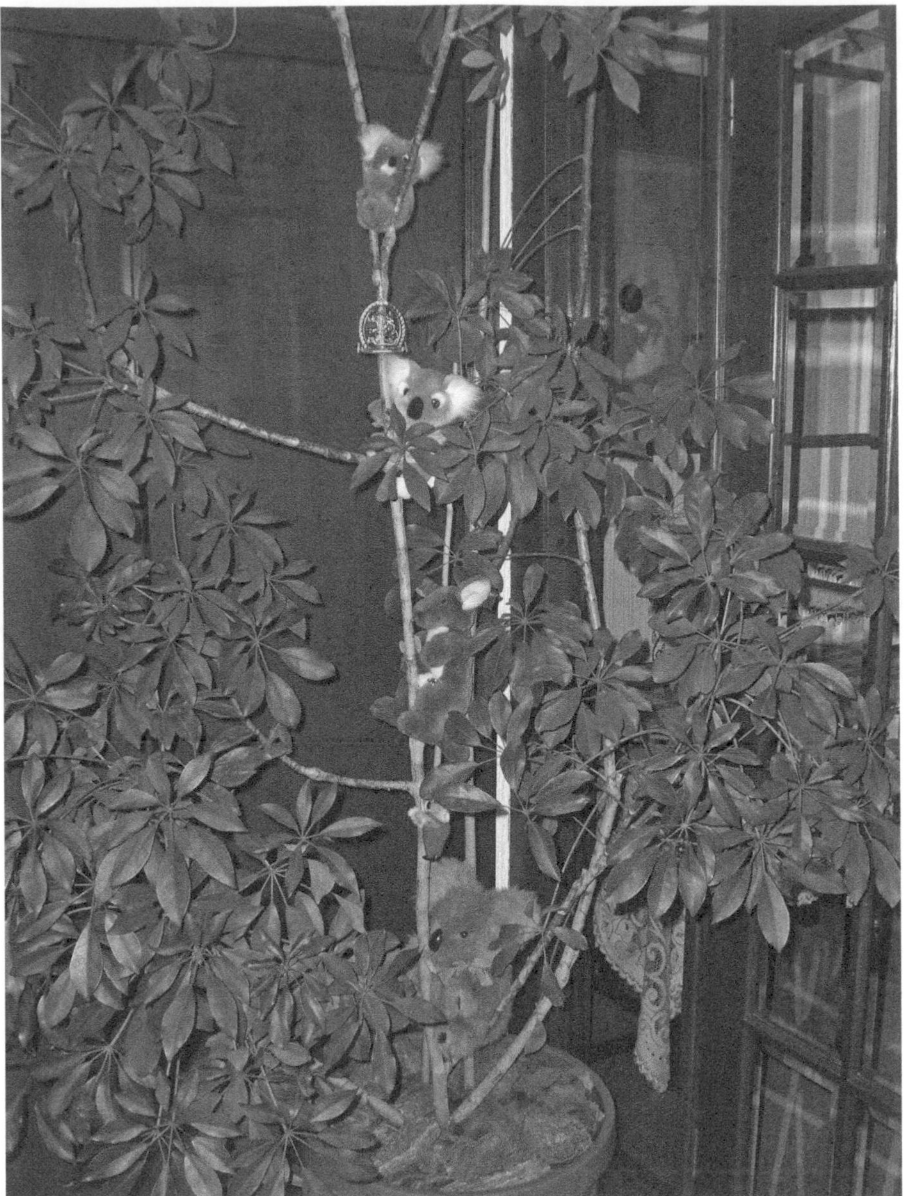

Photograph 7.4 This photograph illustrate the hosts' relationship with a succession of Australian guests

Source: Majella Sweeney

encounter a foreign space to which the host orientates their movements by directing them to bedrooms and other areas that have been reserved for them.

The personal nature of the encounter may reflect a more empathic service and individual attention than in a larger hotel. However, guests may suffer intrusion into their private lives and behavioural preferences as a result of this and may be unable to retain any sense of anonymity or effectively hide their activities from the host's voyeuristic gaze. The level of interest in guests' activities reflects the ways in which they are subjected to processes of 'othering'. The host uses such processes to make decisions regarding social control based on their own definitions of social acceptance and personally defined norms of behaviour:

> If we don't like the look of someone, you know if they misbehave or act strangely then we just won't let them stay. Well, this is a business but its also our home. Of course I want to welcome you here and I'll open my home up to you but I won't have any problems or unacceptable behaviour… I just won't.
>
> (Female host – Di Domenico, 2003)

Derrida (2000: 4) interpreting Kant (Reiss, 1970) draws attention to universal 'laws of hospitality' as well as the 'law of the household'. Whilst universal, or at least assimilated, cultural precepts of hospitality may exist from a guest perspective, the commercial home host may operate idiosyncratic laws of inclusion or exclusion. Derrida (2000: 14) suggests that:

> There is no hospitable house. There is no house without doors and windows. But as soon as there are a door and windows, it means that someone has the key to them and consequently controls the conditions of hospitality.

Despite the existence of national laws requiring hospitality establishments to provide available accommodation on demand to guests willing and able to pay for it, commercial home hosts tend to view it as their prerogative to exclude individuals who do not meet their own definitions of social acceptance, although this is usually orchestrated on an implicit basis due to fear of reprisals from tourist boards or other quality assurance bodies (Jamieson, 2006). Needless to say, such bases of exclusion vary from host to host, and such decisions are inherently mobile within the commercial home. Paradoxically, for the duration of the stay in the commercial home, the guest cannot escape the confines of that home and the host's pervasive personality, behaviours and rules. Instead of representing a liminal space like that of the hotel (Pritchard and Morgan, 2006), where the guest is free to enact whatever performances s/he wishes, the commercial home guest is required to enact performances acceptable to the host.

Consequently, the social control and management of space within the setting of the commercial home is inherently complex due to its multiple functions. Tensions, dilemmas and confusion may arise due to the competing pressures of retaining a sense of home and privacy whilst also providing a sense of freedom and service for the paying guest.

Conclusion: Moving On and Staying Home

Simmel (1970) draws attention to the short-lived nature of human interactions owing to human mobility. Hotels are in many ways liminal non-places (Augé, 1995) providing little opportunity for human interactions. The commercial home by contrast, particularly within the context of short stays, seemingly promises more intensive social interactions, thus enabling travellers to maintain a richer human contact whilst on the move. Yet, the host-guest dynamic is an unequal mobile relationship given the immobile nature of the host and the mobile nature of the guest. However, the presence of guests who bring with them aspects of another place expressed through, for example, language, accent, clothing and their interests, permits a form of vicarious travel. Whilst this is the case, their presence also creates or reinforces a certain sense of the host's immobility. In a world characterized by the importance of mobility such a sense of immobility may be socially significant and despite the pretence of a certain lifestyle, subtly reinforces social divisions regarding subservience of the server and the served.

The commercial home stay may be likened to a form of specialized play, to the concept of an adventure as described by Simmel (1971) in that it affords a differentiation within life leading to the unexpected (Kjolsorod, 2003). It is discontinuous from everyday life but often also continuous as there is the paradox we feel of the strange familiarity of being in someone else's home. This sense of adventure is accentuated by the knife-edge relationship of hospitality which can so easily spill over into hostility (Derrida, 2000; Sheringham and Daruwalla, 2007). In a purely commercial context, the provision of hospitality is always quantified whereas in the commercial home context, the boundaries of commercial hospitality are less clear.

Whilst it may be possible to refer to broad universal laws of hospitality (Derrida, 2000) in the commercial home context, the interpretation of such 'laws' is idiosyncratic, typically as hosts oscillate between commercial and non-commercial interpretations. For the guest, such mobile interpretations of the universal laws of hospitality become a focus of attention, interest or even amusement insofar as they, like their hosts, continuously negotiate their responses to such manifestations in order to endeavour to set out on the journey into the commercial home.

This space of the commercial home beckons to our instincts of homeliness, familiarity, stability and the desire for a genuine welcome. Yet the experience may also prove bitter-sweet as the visitor must negotiate a passage through the personalities, personal possessions, routines and norms of its more permanent residents, and resist any yearnings for anonymity or solitude that the space may not allow. These contradictions, familiarities and peculiarities also affect the host's own voyage, when coming into contact with new faces and ever mobile strangers.

References

Ateljevic, J. and Doorne, S. (2000), '"Staying Within the Fence": Lifestyle Entrepreneurship in Tourism', *Journal of Sustainable Tourism*, **8**(5), 378–392.

Andriotis, K. (2002), 'Scale of Hospitality Firms and Local Economic Development: Evidence from Crete', *Tourism Management*, **23**(4), 333–341.

Aramberri, J. (2001), 'The Host Should Get Lost: Paradigms in the Tourism Theory', *Annals of Tourism Research*, **28**(3), 738–761.

Augé, M. (1995), *Non-places* (London: Verso Books).

Bauman, Z. (1993), *Postmodern Ethics* (London: Routledge Books).

Bell, C. and Newby, H. (1976), 'Communion, Communalism, Class and Community Action: The Sources of New Urban Politics' in Herbert and Johnston (eds).

Bell, D. (2004), 'Taste and Space: Eating Out in the City Today' in Sloan (eds).

Clifford, J. (1997), *Routes* (Cambridge, MA: Harvard University Press).

Cole, S. (2007), 'Hospitality and Tourism in Ngadha: An Ethnographic Exploration' in Lashley, Lynch and Morrison (eds).

Derrida, J. (2000), 'Hostipality', *Angelaki*, **5**(3), 3–18.

Di Domenico, M. (2003), 'Lifestyle Entrepreneurs in the Hospitality Sector'. PhD dissertation (Glasgow: University of Strathclyde).

Di Domenico, M. and Lynch, P.A. (2007), 'Host/Guest Encounters in the Commercial Home', *Leisure Studies*, **26**.

Douglas, M. (1991), 'A Kind of Space', *Social Research*, **59**(1), 287–307.

Ferguson, D. and Gregory, T. (1999), *The Participation of Local Communities in Tourism: A Study of Bed and Breakfasts in Private Homes in London* (London: Tourism Concern).

Haldrup, M. and Larsen, J. (2006), 'Material Cultures of Tourism', *Leisure Studies*, **25**(3), 275–289.

Herbert, D. and Johnston, R., eds (1976), *Social Areas in Cities*, Vol. 2 (Chichester: Wiley Books).

Jamieson, A. (2006), 'Ministers to Stop Scottish B&Bs Banning Gay Couples', *Scotsman*, 14 March 2006. <http://news.scotsman.com/topics.cfm?tid+294&id=379462006> (accessed: 27 .03.06).

Kesselring, S. (2006), 'Pioneering Mobilities: New Patterns of Movement and Motility in a Mobile World', *Environment and Planning A*, **38**, 269–279.

Kjolsorod, L. (2003), 'Adventure Revisited: On Structure and Metaphor in Specialized Play', *Sociology*, **47**(3), 459–476.

Kontogeorgopoulos, N. (1998), 'Accommodation Employment Patterns and Opportunities', *Annals of Tourism Research*, **25**(2), 314–339.

Lashley, C. (2000), 'Towards a Theoretical Understanding' in Lashley and Morrison (eds).

Lashley, C. and Morrison, A., eds (2000), *In Search of Hospitality: Theoretical Perspectives and Debates* (Oxford: Butterworth-Heinemann Books).

Lashley, C., Lynch, P.A. and Morrison, A. (2007), 'Ways of Knowing Hospitality' in Lashley, Lynch and Morrison (eds).

——, eds (2007), *Hospitality: A Social Lens* (Oxford: Elsevier Books).

Lavie, S., Narayan, K. and Rosaldo, R., eds (1993), *Creativity/Anthropology* (Ithaca, NY: Cornell University Press Books).

Lowe, A. (1988), 'Small Hotel Survival: An Inductive Approach', *International Journal of Hospitality Management*, **7**, 197–223.

Lynch, P.A. (1998), 'Female Entrepreneurs in the Host Family Sector: Key Motivations and Socio- Economic Variables', *International Journal of Hospitality Management*, **17**, 319–342.

—— (2003) 'Conceptual Relationships Between Hospitality and Space in the Home-Stay Sector'. PhD dissertation (Edinburgh: Queen Margaret University College).

—— (2005a), 'Sociological Impressionism in a Hospitality Context', *Annals of Tourism Research*, **32**(3), 527–548.

—— (2005b), 'The Commercial Home Enterprise and Host: A United Kingdom Perspective', *International Journal of Hospitality Management*, **24**(4),

Lynch, P.A., McIntosh, A. and Tucker, H., eds (2007 forthcoming), *Commercial Homes in Tourism: An International Perspective* (London: Routledge Books).

Madigan, R. and Munro, M. (1991), 'Gender, House and "Home": Social Meanings and Domestic Architecture in Britain', *Journal of Architectural Planning and Research*, **8**(2), 116–132.

Marcus, C.C. (1995), *House as a Mirror of Self* (Berkeley, CA: Conari Press).

Morrison, A. (2002), 'Small Hospitality Businesses: Enduring or Endangered?', *Journal of Hospitality and Tourism Management*, **9**(1), 1–11.

Myerhoff, B. (1993), 'Pilgrimage to Meron: Inner and Outer Peregrinations' in Lavie, Narayan and Rosaldo (eds).

Osborne, P. (2000), *Traveling Light: Photography, Travel and Visual Culture* (Manchester: Manchester University Press Books).

Pritchard, A. and Morgan, N. (2006), 'Hotel Babylon?, Exploring Hotels as Liminal Sites of Transition and Transgression', *Tourism Management*, **27**, 762–772.

Reiss, H., ed. (1970), *Kant's Political Writings* (Cambridge: Cambridge University Press).

Ritzer, G. (1996), *The McDonaldization of Society* (London: Pine Forge Press).

Shaw, G. and Williams, A. (2004), 'From Lifestyle Consumption to Lifestyle Production' in Thomas (eds).

Sheller, M. and Urry, J. (2003), 'Mobile Transformations of "Public" and "Private" Life', *Theory, Culture & Society*, **20**(3), 107–125.

—— (2006), 'The New Mobilities Paradigm', *Environment and Planning A*, **38**, 207–226.

Sheringham, C. and Daruwalla, P. (2007), 'Transgressing Hospitality: Polarities and Disordered Relationships?' in Lashley, Lynch and Morrison (eds).

Simmel, G. (1970), *The Sociology of Georg Simmel.* Wolff, K. (ed.) (New York: The Free Press).

—— (1971), *On Individuality and Social Forms*. Levine, D.N. (ed.) (London and Chicago: University of Chicago Press Books)

Sloan, D. (2004), *Culinary Taste: Consumer Behaviour in the International Restaurant Sector* (Oxford: Elsevier Books).

Sweeney, M. (ongoing research) the *Host's Relationship with their Commercial Home*, Ongoing Doctoral Research (Edinburgh: Queen Margaret University).

Sweeney, M. and Lynch, P.A. (2007), 'Explorations of the Host's Relationship to the Commercial Home', *Tourism and Hospitality Research* 7(2), 100–108.

Telfer, E. (2000), 'The Philosophy of Hospitableness' in Lashley and Morrison (eds).

Thomas, R., ed. (1998), *Small Firms in Tourism: International Perspectives* (London: Elsevier Books).

Treadwell, S. (2005) 'An Image of Elsewhere', *Space and Culture*, **8**(2), 214–224.

Tucker, H. and Keen, D. (2005), *Hosting Guests in Rural New Zealand: A Social Analysis of the Bed and Breakfast and Homestay Sector* (Dunedin: Department of Tourism, University of Otago).

Tucker, H. and Lynch, P.A. (2004), 'Host-Guest Dating: The Potential of Improving the Customer Experience Through Host-Guest Psychographic Matching', *Journal of Quality Assurance in Hospitality and Tourism*, **5**(2/3/4), 11–32.

Urry, J. (2000), *Sociology Beyond Societies: Mobilities for the Twenty-First Century* (London: Routledge Books).

Walton, J. (1978), *The Blackpool Landlady – A Social History* (Manchester: Manchester University Press Books).

Ghekko **(the Last Resort, Ko Lanta, Thailand, 2004) by Elly Clarke**

Chapter 8

Hospitality in Flames: Queer Immigrants and Melancholic Be/longing[1]

Adi Kuntsman

This chapter examines the ways different ideas about hospitality are mobilized to constitute queer migrant subjectivity. In particular it addresses the ways immigrants' sense of belonging works through their relations to a community space, and how this space, in turn, constitutes immigrants as mobile and/or rooted in the past and in the present. The chapter is based on my ethnography of the Russian-speaking queer community in Israel.[2]

Since 1989 more than a million people have immigrated to Israel from the former Soviet Union, welcomed by the Israeli 'Law of Return'. Russian-speaking immigrants occupy an ambivalent position in Israeli society. As Jews they are welcomed by the Zionist ethos of national homecoming, but at the same time they are stigmatized as racially and culturally 'impure' immigrants that must be turned into proper Israelis. Ethnically, Russian-speaking immigrants are *Ashkenazi* (of European origin), light skinned, and many among them have higher education – as such, they are desirable for the Zionist project of Israeliness as white middle-classness and Europeanness. And yet they are often seen as not being of the right kind of Europeans, coming from supposedly backward and undemocratic Eastern Europe.

In the last 15 years the 'Russians' (as they were often called) have been navigating their place in Israeli society, moving between integration and separate ethnic infrastructures. They created a significant public presence including Russian language newspapers, shops and community centres. Around 1999–2000 the gay,

---

1 I would like to thank Hadassah-Brandeis Institute on Jewish Women for contributing towards fieldwork and research expenses. I am grateful to my supervisors, Dr Anne-Marie Fortier and Dr Gail Lewis for their comments on earlier versions of this paper. I thank the editors, Jennie Germann Molz and Sarah Gibson for their comments on the earlier draft and for the guidance through the revision process. And last but not least, I am indebted to my partner, Yehudit Keshet for her editorial assistance with this paper and for her overall invaluable help and support.

2 This chapter is part of a 10-month-long ethnography that included participant observations on the website, online and face-to-face interviews and archival analysis of hundreds of discussion threads over a period of 2.5 years. I received permission from the website's two co-administrator to use the archives. Other participants have been informed about my research.

lesbian, bisexuals and transgendered (GLBT) among them also began organizing.[3] Today the Israeli queer scene also has its 'Russian' spaces. The most notable among them are *At Roby's* – the first, and so far, the only club for Russian-speaking queers, and an online discussion forum for Russian-speaking gays, lesbians, bisexuals and transgenders. At the time of my research in 2003–2004 the club and the forum were the two main sites of 'Russian' queer presence, closely linked to each other. Many of the forum's participants ended up meeting at the club. The forum was also a space to discuss the club itself, express opinions and criticize it. In these discussions the club was depicted as either an ultimate place of welcome or as a site of failed hospitality. Such discussions often turned into violent fights, known in cybercultures as flame wars, in which people behaved aggressively, insulted, cursed, name-called and mocked each other.[4]

Although the two sites are closely related, the discussion of these relations lies beyond the scope of this chapter. What it does address is the notion of hospitality as it arises from the flame wars about the club. As this chapter will show, these fights embody the ambivalence of the club's visitors, who are simultaneously welcomed and othered both in Israeli society as a whole and on the local queer scene. As first-generation immigrants in their twenties and thirties, Russian-speaking queers are torn between the Russian past and the Israeli present, between longing for a place of their own and a desire not to be seen as different from Israelis. Their visions of hospitality embody the complexity of collective identification as immigrants and as queers. Any hospitality for them is far from taken for granted; rather, it is a site of struggle, splitting and impossibility.

At Roby's – A 'Place of Our Own'

In 2001, Roby, an immigrant woman in her early 50s, began organizing 'Russian evenings' for women at Tel Aviv's lesbian feminist centre. I remember talking to Roby then. 'I don't want to ask anything from them', she said, meaning the Israeli women who run the centre. 'I want a place of our own.' Soon she indeed opened a small club in a private apartment in South Tel Aviv. After another year *At Roby's* moved to a bigger venue, still in South Tel Aviv, where the club is now permanently located.

Roby's remark struck me with its pain and longing, and also with its sense of alienation from – and disidentification with – the Israeli scene. A desire for a 'place of our own' is a familiar experience for many immigrants. Located in different geo-political contexts, some of them leave a national home behind, others arrive at a 'new home', and yet others move between different locations simultaneously coded as 'homes'. Some have a place (or places) they can call 'home'. Others, not having that luxury, still long for – and actively establish – places of attachment, driven by what Avtar Brah calls 'homing desires' (1996:192). She emphasizes that homing

3 Until then most of the queer immigrants were either isolated or involved in the Hebrew-speaking LGBT community.

4 Around one-third of all the discussion threads about the club that I collected could be characterized as flame wars.

desires are not the same as a desire for a national homeland, and are not necessarily linked to a fixed place of origin. Home, for Brah, is first and foremost a 'mythic place of desire in the diasporic imagination' (1996: 192).

The notion of homing desires has also become a familiar trope in queer studies that particularly focused on 'coming out' to queer culture as 'coming home' (Fortier, 2001). Scholars of 'queer diaspora', working at the intersection of sexuality, migration and mobility, have approached this idea critically. They have pointed out that many queer communities do not carry the same promise of belonging for all gays and lesbians (Manalansan, 2003; Petzen, 2004), alienating, excluding or ignoring gays and lesbians of colour. At the same time they note that for many diasporic and migrant queers a desire for places of attachment and belonging is crucial. Critical of the linear narrative of moving into a (white) queer home/subculture, they nevertheless emphasize the importance of homely places and homing desires in queer immigrants' lives, shifting the discussion from a register of travel/movement to one of affect/attachment (Fortier, 2001).

Informed by these debates, I explore the forms of attachment and homing in the context of mobility. What kind of dreams about hospitality and welcome arise among immigrants who left their country of birth and moved into what is defined by the state as their national home? How do the immigrants imagine themselves as guests, hosted by other immigrants? What is a queer immigrants' 'place of our own'? And last but not least, how can the queer scene become a place of disidentification and loss?

Hospitality for the Soul

Many participants describe *At Roby's* as a place for the soul, a place that welcomes them in a particular way. Explaining his need for a 'Russian' club, one participant, Boris, wrote:

> We see Israelis in our own way. They see us in their own way. The older generation of immigrants doesn't see GLBT the way Israelis do. We, as Russian-born or as their descendants, naturally have something special in the breadth of our soul and our way of relaxation. Roby's club warms up this particular part of our soul. That's why it exists.

In his poetic description of naturalized cultural differences, Boris narrates the club as a place for his particular needs as an immigrant: to relax his soul, and also to *be*, rather than to be seen (as different). The idea of hospitality for the soul has its roots in the Russian tradition of welcoming and turning strangers into guests by opening for them the doors of one's home. Dale Pesmen, in her analysis of 'soul' in Russia , notes that 'a common description of hospitality implied that a host or hostess's job was to create a substitute home' for the guest (2000: 161). Being welcomed and made to feel like at home is, too, a common trope in some postings about *At Roby's*, where it is used to emphasize the club's uniqueness. 'In my opinion', wrote one participant, Alex, 'there is something about this place that makes me feel like at home.... This feeling, I must tell you, is worth a fortune. I have been to many gay places, but have never felt as free [comfortable] anywhere else.'

Alex describes the club as a place where his homing desires are met. For him this is first and foremost a place of good feelings: he feels at home and relaxed. The expression he uses – to feel free/relaxed/comfortable – is in fact best translated into English as 'feeling at home'. Alex feels at home because he knows the club's owners. But the idea of feeling 'free', as opposed to feeling constrained, tense and self-aware, is also linked to the Russian language. As several people noted, one of the things that attracts them to '*At Roby's*' is the ability to communicate in their mother tongue, while also being in a queer place. For Alex , *At Roby's* club appears as an idealized place of welcome. But it is not just a place where traditional Russian laws of hospitality operate (as opposed to Israeli clubs, that do not have – and do not welcome – the soul); it is also a particularly precious place of immigrant belonging, where one can feel at home as a 'Russian', as an immigrant and as queer. Importantly, this sense of belonging is non-material, or more specifically, it is juxtaposed to the commercial logic that underpins most of the queer scene, in Tel Aviv and many Western urban areas. The sense of connectedness cannot be bought, many participants emphasize. In a discussion over the quality of food at the club, one person noted: 'not by bread alone... Even the best and tastiest steak cannot compare to [the pleasure of] sitting with friends over a beer (although I personally love good food)'.

'Not by bread alone' refers to the commonly known quote from the Bible: 'Man shall not live by bread alone, but by every word that proceeded out of the mouth of God'.[5] Rooted in Judeo-Christian tradition, widely known and commonly used in Russia, the phrase does not necessarily carry a religious connotation but rather refers to the secular-spiritual emphasis on the non-material, which in turn is seen as a sign of having a soul. The reference to the Judeo-Christian notion of spiritual life in the context of drinking might appear as puzzling, after all, drinking, just like eating, is a bodily act. But Snake's text would better be understood within the role 'sitting' and drinking has traditionally played in Russian and Soviet society, where it was a highly ritualized act, loaded with social meanings. Drinking and sitting together, emphasizes Pesmen in her ethnography (2000), is a genre of communication central to Russianness. 'Rituals including alcohol were the epitome of hospitality, condensing economic, sociocultural, philosophical and psychological *dusha* [soul, Russian]. [...] Drinking situations [...] were also privileged contexts for other *dusha*-related activities, such as playing music, singing, cursing and reciting poetry' (2000: 171). Drinking together, continues Pesmen, can act as a communion, and as a way of establishing the boundaries of terms like 'friends' or 'one of us'. Similarly, in Snake's posting, sitting and drinking make the club a place for the soul. The club becomes a place of the immigrant queers not only because it targets a particular group in the population, but because of the form of communication that it allows.

Images of welcome and belonging, evoked by Snake and others presented in this section, position the club as what Svetlana Boym calls 'diasporic intimacy': a form

---

5 These words appear twice, first in the Old Testament (spoken by Moses to the Israelites in the Exodus), and in the New Testament (Matt. 4:3–4).

of connectedness between immigrants, as 'a precarious cosiness of a foreign home' (2001: 254). I read Boym's reference to diasporic intimacy as a form of attachment *in* mobility, a form of belonging as well as uncertainty. Indeed, the connection between the club's visitors is at the same time deeply rooted in the Russian heritage and fragile because of its status as a marginal, immigrant space on the Israeli queer scene.

But not all the participants agree on the vision of *At Roby's* as a place they can or want to consider their own. Some see it as a locus of failed hospitality, as a repulsive, disgusting and unwelcoming place, and as a concentration of everything they do not want to be and do not want to be associated with. For them it is a place they want to move away *from* rather than a place they long to come to. Their descriptions of the club are written in poisonous and sarcastic language; in some of them the club's owners, employees and visitors are described in derogatory terms, and even openly insulted. But most crucially, the critique of the club assaults the very foundations of hospitality for the soul, and at times, it seems, they are specifically attacking it. Together with it, they also attack the Russianness of the club, or rather, mark the club as a particularly Russian queerness they want to set on fire.

Failed Hospitality

One of the most frequent critiques of the club was that it failed to provide appropriate food and drinks. Strikingly, the many complaints about food and drinks depicted them as specifically 'Russian' or Soviet, and therefore as not good. One participant wrote that the food offered at the club did not look appealing and reminded him of a Soviet public dining centre. Soviet dining centres were known for their bad food, and can be also seen as an antonym of welcome, both because of the poor cooking and because these were very public spaces, subjected to surveillance and control, and opposed to the intimacy of informal communication.

Drinking at the club, too, was depicted in a negative way. Many participants noted that they were repulsed by the drunkenness of the club's visitors and some have graphically described their encounters with people puking in or outside the club. Their feelings signalled disidentification: not only that they did not want to be associated with these other visitors, but they also repeatedly created distance between themselves and others through expressions of disgust. Disgust, as several feminist scholars have noted, is a powerful emotion often used to create a boundary of belonging (Probyn, 2000; Ahmed, 2004). Sara Ahmed, for example, emphasizes the performative power of disgust to constitute both the disgusting object and the community of those disgusted by it (2004). The act of disgust – expelling the abject – is the opposite of the act of hospitality as taking *in*; being disgusted, in other words, can be seen as the opposite of being welcoming.

But disgust, notes Ahmed, is also a way of 'sticking' objects together through metonymic proximity. In the many descriptions of the club as disgusting two interrelated themes are particularly interesting in that respect: the classed character of drinking and the 'Sovietness' of the club. Many of those disgusted by the club do not see drinking as a 'spiritual' connection with friends, but as 'Russian' drunkenness that contrasted negatively with knowledgeable drinking. This contrast constituted drinking as a form of cultural capital (Bourdieu, 1984). One of the critics, Crazy Ala, wrote:

> We ordered Whisky-Cola but it turned out that there isn't enough whisky for the second portion ... We were offered "Sir Kent" which we decided not to drink for hygienic reasons. At the bar [people were] drinking beer, probably "Baltic"[6] ... A woman at the bar could not figure out the price for the whisky and called from the back room for a woman named Olga, whose drunken mug confirmed the cult status of this place.... We decided not to waste money on dubious alcohol, got a taxi and went to "Carpe Diem" where we had a few high quality cocktails that somewhat improved the bad taste left by *At Roby's*.

Crazy Ala appears here as a sophisticated drinker. When her demands are not satisfied, she and her companion leave for the fancy and Bohemian Tel Aviv gay bar 'Carpe Diem'. But who are the people at the club they leave behind, having mocked its 'cult' status? Those at the club drink dubious alcohol: cheap vodka and Soviet beer. They drink without differentiation, in order to get drunk. The woman at the bar – a 'drunken mug' – comes to represent all visitors. Her ugly appearance is clearly associated with Russian images of alcoholism, and constitutes the opposition between knowledgeable drinking and ugly getting drunk. She marks the whole place as a site of unsophisticated and unpleasant alcohol consumption. The drunkenness is repeatedly mentioned in conjunction with club's 'Sovietness', metonymically linking the two to each other and to the immigrants' past, the past they want to move away from.

The depiction of the club and its visitors as immigrants stuck in time comes back again and again in the critique of the club. One frequently repeated theme in the complaints about the club was that it looks like the Soviet Union of the 1980s. 'What is the club rated for? Bad taste and provinciality?', wrote another participant. 'When I see it I feel that the time stopped in 1990'. Another one noted: 'When I go to a club, I come to dance, surrounded by on the ball, good-looking people. I don't want to languish in a corner with a plastic cup, watching what looks like a disco in Zhitomir in 1987'. Zhitomir is the name of a small town in Ukraine, known for its rich Jewish life in the late-nineteenth and twentieth centuries. As one of the places with large Jewish communities, it can be characterized as *shtetl*, a Jewish village/town, one of many that existed beyond the 'Pale of Settlement'[7] during the tsarist regimes, and remained after the revolution. But in pre- and Soviet-Jewish mapping, *shtetl* is also a marker of provinciality, the opposite of the assimilated Jews who live in big cities. Here Zhitomir figures as a sign of traditional, stuck-in-time Jewish as well as Soviet provinciality.

But the Sovietness of the club is also the opposite of modernity, in particular, queer modernity. It is not accidental that the Soviet Union always appears as a moment frozen in time: the late 1980s. One thing that comes to mind is of course the

---

6 'Baltic' is the name of a Soviet beer, one of the few kinds that existed before the beginning of mass exports from abroad in the late 1980s.

7 Jews in the Tsarist Russia were only allowed to settle beyond the 'Pale of Settlement', in rural areas and not in big cities.

collapse of the Communist regime that began with the destruction of Berlin Wall and was soon followed by the fall of the Soviet Union itself. Late 1980s, for the West, is probably registered as the last days of the Soviet Empire. But for the Russian-speaking immigrants of the last wave, and in particular, for the immigrants in their late twenties and thirties, the late 1980s also carries a deep personal and generational significance. For them the Soviet 1980s is the time (and place) of their adolescence or early adulthood. It was also the last moment of their lives in a country that no longer exists. It became fetishized and frozen in their memory, living through stereotypes, through moments of longing but also through shame and disidentification. So while some of the club's visitors may embrace the nostalgic pleasures of listening to pop music or watching a drag performance that involves songs and personae of their childhood, others refuse this association and instead want to merge with the shiny and up-to-date Israeli scene, as they perceive it. This scene in itself – in its mixture of multicultural consumption, racist and classist exclusions, its fantasies of Western-ness and depoliticized sexuality – is a complex and contradictory cultural site. But in the postings that criticize *At Roby's* and glorify Israeli gay clubs, the Israeli scene appears as an idealized place, not burdened by a past, and welcoming for everyone.

Tellingly, this is exactly the vision that arises from the descriptions of *At Roby's* that appeared in two mainstream Israeli GLBT magazines.[8] The two articles portrayed the Russian club with a mix of disdain, exoticization and excitement. One journalist described the club as located at the margins of queer Tel Aviv: 'outside the pretentious posturing of the (gay/night) scene, alongside the straight clubs of South Tel Aviv'. The outsider status of *At Roby's* is repeated twice: once when it is metaphorically located beyond the Israeli queer scene, and then when it is geographically placed in South Tel Aviv – the area of slums, immigrants and urban poor. 'Do you want to feel as if you're abroad, in another country? Come to the place called *At Roby's*,' wrote the other journalist, informing its readers that 'with the assimilation of "Russians" in Israeli society, it was decided to open the club's doors for those born in the country, who were thirsty for new experiences' (Rinat Haimov, 'Abroad at Roby's', *Ha Zman Ha Varod*).

Both articles list the stereotypes of the Russians/Soviets: blond curls, blue eyes, Kazachok, 'round-dancing' and athletic bodies. These stereotypes resemble Hollywood-style representations of Soviet Russia. One of the journalists writing about *At Roby's* even demonstrated some 'knowledge' of Russian, to make her description look more authentic: she completes her article with 'Fashes darovie, comrades'. This is a distortion of 'za vashe zdorovje' – 'to your life', a Russian toast that appears in almost every cinematic representations of Russians. And of course, both articles mentioned the music of the 1980s. The frozen Soviet past and the white Siberian Russianness seem to haunt the club and its visitors. But these stereotypes – and in particular the blond hair and blue eyes – also painfully resonate with the way Russian-speaking immigrants are sometimes depicted in the Israeli media as 'too white', that is, as non-Jewish (Golden, 2003) or not Jewish enough. Within the framework of their immigration as national homecoming, and within the

---

8 Oren Ziv, *TOP Magazine* and Rinat Haimov, *Ha Zman Ha Varod,* in 'At Roby's' [website], <http://www.aguda.org/community/shunya.php>, accessed May 2004.

Israeli economy of Jewishness, blond hair and blue eyes became the synecdoche for the Russian-speaking immigration, marking their desirability as Europeans and at the same time their marginality as racially mixed and 'Russian' rather than Jewish. Russian-speaking immigrants were often seen as 'failing' Jews who have not maintained blood purity and lost their Jewishness to numerous mixed marriages, as well as losing their cultural and religious identity due to assimilation in the diaspora. Not surprisingly, the club is described as 'another country', forever marked with difference despite 'assimilation', this time in Israel.

The two articles were translated and published on the site. Some people laughed them off, others were clearly upset. 'This is all they see in us, Russians,' wrote Will, 'That we drink vodka non-stop and dance "Kazachok"'.

But how do they see themselves? What does 'Russianness' mean for Jewish immigrants, whose identification with Russia is complicated and ambivalent? For example, the figure of the Russian drunk has often been seen as the antipode of the (Russian) Jewish intelligentsia, both in terms of class and drinking practices (Lomsky-Feder and Rapoport, 2003). Figuring the club as a place of drunkenness juxtaposed to fashionable drinking is a performance of classed queerness; at the same time, it can undoubtly be read as internalization of stereotypes of 'the Russians', circulating in Israel and in the West. But the disidentification with the club because of its drinking culture is also a form of claiming one's Jewishness, against the 'Russianization' of immigrants by Israeli society. As many of the Russian-speaking immigrants in Israel repeatedly noted, referring to their racialization, 'we were Jews in Russia and became Russians in Israel'. They emphasize the irony of finally becoming 'Russians' in the country that welcomed them as Jews, after having been rejected from Russianness in the Soviet Union. Rejecting the club's Russianness is therefore not only about the club's hospitality, but about immigrants' sense of belonging in Israel. Being welcomed *At Roby's*, in other words, expresses the desire to be welcomed in the country;[9] the hospitality of the queer club is burdened with the contradictions that are part of the immigrants' daily experience in the past and in the present. This ambivalence of desire, rejection and (dis)identification informs the image of the club as welcoming and disgusting, as too Jewish (provincial, from a *shtetl*) and too Russian (full of drunks). It is not surprising, that the image of *At Roby's* as a place of diasporic intimacy and hospitality for the soul so often coexists with the feeling of impossibility and loss.

At Roby's – A Place of Loss

Those who criticized the club often approached it through the lens of commercial hospitality: an impersonal but efficient service for one's money (as opposed to Russian hospitality for the soul). 'I don't care about Roby's soul, I just want to be served', wrote one of the participants, Richie, in response to the description of *At*

9 One must never forget, of course, that the welcoming of Jewish 'homecomers' into Israel is based on and goes hand in hand with the expulsion of Palestinians from their lands and homes. The relations between the Israeli colonial project and immigrants' sense of welcome deserves a separate discussion. See, for example, Kuntsman (2006).

Roby's as a place for the soul, 'Roby doesn't listen to her clients and loses their trust'. But is it her loss, or theirs? It seems, that in many passionate attacks on the club the one who feels loss is, in fact, the complainant him/herself.

> Nice Tree: 'Push off to *Roby's*'
>
> Friday night, the eternal search for entertainment... Where should we go? For the Russian gay crowd there isn't much choice – all roads lead to *At Roby's*...
>
> I remember the old place[10] where everything was nice, and the music was good, although the place was tiny. My last visit to the place brought only negative impressions. The music at the club is such a mishmash that you don't want to dance at all. 'First we need to have vodka', I heard two women saying when entering the club. Indeed, if you want to stay at this place for even 5 minutes you must be drunk...
>
> And the window display [that looks like] a Russian shop with all kinds of products. Grilled meat from the morning, reddish, and hell knows what else... looks not too appealing. So what is it? Reminds me of a Soviet [public] dining center of the late 1980s. And the prices at the bar are so expensive they scare me away... And the people are no longer the same... Well, here I won't say a word, you know it yourself. The toilet was inaccessible because some boy full of make-up was puking into the sink...
>
> I felt so sad, and afraid. Its 2004 now, but feels like 1989. Roby's lack of taste and business initiative makes the place go from bad to worse. Of course, the administrator will erase my terrible thoughts, but I am posting them not to hurt anyone, I just want Roby to take this into account... Life is a complicated thing.

Nice Tree's complaints about the club (as well as many other posting criticizing *At Roby's*) are written in a genre that Nancy Ries (1997) characterizes as 'litany'. In her work on distinctive Russian speech genres, Ries describes litany as

> passages in conversation in which a speaker would enunciate a series of complaints, grievances, or worries about problems, troubles, afflictions, tribulations, or losses, and then often comment on these enumerations with a poignant rhetorical question ('Why is everything so bad with us?'), a weeping, fatalistic lament about the hopelessness of the situation, or an expressive Russian sigh of disappointment and resignation (1997: 84).

Indeed, many complaints about *At Roby's* include the fatalistic elements of lament about the fate of gays in general and of Russian gays in Israel, in particular. Nice Tree's posting, in particular, is a story of loss. He used to like the club in its early days, but these days – and the old club – are now gone. Instead he finds himself in a place where everything is bad: his posting is yet another outcry about the drunks, the puke, the food, the music and the sense of being stuck in time. Nice Tree's mourning, emphasized by his 'I feel so sad, and afraid', and the sense of being trapped in the wrong kind Russian queerness ('All roads lead to *At Roby's*') resonate with many other postings that criticize the club. The unsatisfied, sad and disgusted visitors were looking for a different Russian queerness which they could not find at the club. Their

10 One year after opening, the club moved to another venue.

hopes turned into loss, and their loss, in turn, became an attack. For some, like Nice Tree, it was a loss of the place that they once had and liked. But for many others the Russian queer 'place of our own' was never there in the first place, at least not in the way they wanted it to be, and they mourned its impossibility. One participant, Spring Breeze, wrote a long posting about her desire for a Russian-speaking queer scene and for intimacy she can only find in her mother tongue. She then turned to her negative impression of the club, written in the style of 'poignant rhetorical questions':

> Spring Breeze: I probably have really strange demands of the owners of At Roby's... if I order something non alcoholic, why do I have to wait until the woman serves everyone who came after me? Why can't I drink from a clean glass (I don't even dream about special glasses for wine or juice), why do I get my orange drink in a disposable plastic vessel? And by the way, drink and juice are not the same, so why if I asked for juice, do I get a drink? Why is the change always either sticky or wet?

She concluded her long list of complaints (only partially reproduced here) by exclaiming: 'This is all so sad, and there is no hope for change. That's why I don't come to *At Roby's*, and go to meet women in Hebrew-speaking places, where the unexpected sound of the mother tongue comes like life's unintentional gift'.

Ries points out that litany, widely used by people in Russia in their everyday conversations as well as in public performances in media and politics, constitutes a recognizable Russian stance,[11] a 'posture that expresses particular perspectives, values, desires or expectations' (Ries, 1997: 88). This is often the stance of the victim, the perpetual sufferer, whose suffering gives him/her a position of moral sanctity. At the same time, litanies are an important cultural tool to 'create multiple fields of identification and belonging' (113). Taking Ries' argument further, I read the complaints about *At Roby's* as performative speech acts (Butler, 1990). The repetition of complaints – and the flame wars they generate – constitutes the imagined community of Russian-Israeli queers as a community of suffering. Their suffering is a double-bind: it is embedded in the racial structure of Russian, Soviet and Israeli societies; but it is also self-inflicted. While people keep rejecting the club, writing extensively about the unpleasant atmosphere and dissatisfying service, they seldom walk away. On the contrary, they return to the Forum to weep and to attack. Their flaming words hurt the club and those participants who regard it as a place of their own. But in starting a flame war the flamers also expose themselves to violence: in every thread that contains a complaint about *At Roby's* the attack on the complainer follows immediately. The repetition of the inflammable postings and flame wars functioned similarly to Nice Tree's masochistic move, combining longing with repulsion. The suffering subject always returns to the lost object, to double, rather than resolve, the suffering: it is both the pain of not having a place to go, and the pain of being attacked by others.

---

11 Importantly, Ries (1997) also reminds us that it is easily stereotyped and parodied.

Melancholic Be/Longing

Queer immigrants' homing desires, as this chapter has demonstrated, were always mediated by the notion of hospitality: the club and its owner were seen as a host that was/had to welcome its guests in a particular way. But the flame wars about the club had another model of ownership. Anne-Marie Fortier in her discussion of the Italian community in London proposes to examine ethnic organizations 'as cultural objects, as collective property, as a *scene* for referring to and experiencing collective affective belonging' (Fortier, 2006: 65, original emphasis). Indeed, those who cherish the club as a place for the soul, as well as for those who depict it as disappointing and disgusting, seem to figure *At Roby's* as a collective possession. *A place of our own* reads almost literally: the sense of ownership, it seems, is what allowed so much critique as well as passionate defense of the club. This idea of the club as collective possession contradicts the understanding of hospitality where the host is constituted as having ownership. The immigrants, in that respect, are simultaneously hosts and guests of their club. But what about the fights? Why do the participants constantly attack the club and each other? Why do those repelled by the club keep coming back to the forum to talk about it?

The repetitive and passionate flame wars signal deep ambivalence in the queer immigrants' relations to the club: desire as well as loss, attachment as well as repulsion, welcome as well as rejection. In his discussion of mourning and melancholia, Freud (1917/34) distinguishes between mourning in which the lost object is grieved until the subject can move on; and melancholia, where the grief is not resolved because the subject refuses to let go of the object and thus cannot redirect the affect/desire to another object. Freud further elaborates, that while in the case of mourning the lost object is fully known, in melancholia it is not always clear *what* has been lost: the melancholics themselves do not know what they have lost. For Freud, melancholia, unlike mourning, is mainly an unconscious process that is hard to comprehend and cure. It is this sense of loss of what is not known that can best describe the endless and always returning complaints about the club. Notably, a request for specific suggestions that comes from some participants towards the complainers, almost never brings up a discussion. The intense emotions of the flame wars circulate in lamenting the impossible club, not in constructive proposals for its change. This is not because of an unwillingness to change the club, as some people have suggested. Rather, it results from the ambivalence of Russian-Israeli-queer identification itself which turns the club into an object of melancholia: a loss without an end for the mythic place that was never there.

Of course, the term melancholia should be used cautiously: I do not aim to pathologize the immigrants' longing for a place and their dissatisfaction that generates endless flame wars. My analysis is also in no way a diagnosis of a disorder, whether individual or collective. My deployment of Freud's work differs from the psychoanalytic or therapeutic use of the terms mourning and melancholia: I am not addressing them as 'normal' versus 'pathological' processes and do not read melancholia literally, as an illness (as Freud defines it in 1917).[12] Following Eng and Han, I am interested in melancholia as a 'depathologized structure of feeling' (2003: 344) and as a sociocultural, rather than only an individual, process. In their discussion of race and the American myth of immigration as assimilation, Eng and Han note that for Asian Americans (and other groups of colour) the process of assimilation is suspended and unresolved. Whiteness for them remains 'at an unattainable distance, at once a compelling fantasy and a lost ideal' (2003: 345). Eng and Han's reading of immigration within the melancholic framework is useful for my understanding of Russian-speaking queers' longing for an impossible 'place of their own'. This impossibility, I would like to suggest, is embedded in their ambivalent position as Russian Jews, as Russian-speaking Israelis, and of course, as Russian-speaking queers. They are at the same time mobile and rooted, 'homed' and homeless. For many of them both 'Russianness' in the Soviet Union and Israeliness in Israel remain at an unattainable distance, although of course there are significant differences between the two modes of inclusion and exclusion in these two societies. Roby's club embodies the ambivalence of the immigrants' collective identification as those who were Jews in Russia and Russians in Israel. They want a place to speak Russian, but reject the club's other aspects, seen as manifestations of 'bad' Russianness. The 'Sovietness' of the club is ambivalent, too: it can evoke nostalgia for music and the atmosphere of one's childhood and adolescence. At the same time, however, it signals exclusion from the Israeli queer scene with its 'pretentious' modernity, so the club comes to stand for its visitors' uncomfortable ambivalent position, as homecomers without a home.

'The loss of a love-object', writes Freud, 'constitutes an excellent opportunity for the ambivalence in love-relationships to make itself felt and come to the fore' (1917/34: 161). Indeed, as this chapter demonstrated, the ambivalence of Russian-Israeli queerness comes into the open in the endless fights about *At Roby's*. Reading the immigrants' affective relations to the club as melancholic ambivalence one can better understand the violence of the flame wars about it. Freud notes that melancholic grief and the refusal to let go of the lost object can turn the violence inwards, to the self. He describes it as the 'self-torments' of melancholics, which are both painful and pleasurable: 'If the object-love, which cannot be given up, takes refuge in narcissistic identification, while the object itself is abandoned, then hate is expended upon this new substitute-object, railing at it, depreciating it, making it suffer and deriving sadistic gratification from its suffering' (Freud, 1917/34: 161–162).

'Railing at and depreciating' indeed characterizes the endless flame wars and the ways many participants describe Roby's club and its visitors, whether real, or potential – those who might want to go there. In the postings criticizing the club *At Roby's* appears as a substitute object for the club queer immigrants lost without ever having. Their violence is directed towards the existing club, but also towards those visitors who do not mourn the loss but find what they long for, as queers and as immigrants. But the flame wars, I believe, are also a form of displaced bitterness

12 In his early work on the subject Freud (1917) distinguishes between the 'normal' process of mourning and the 'pathological' melancholy. In his later work he rethinks the relations between the two and suggests that subject formation always includes melancholy.

towards the unwelcoming Israeli queer scene, where the much desired sound of Russian can only be life's unintentional gift and where their own club is seen as another country. Queer immigrants simultaneously desire and resent this Israeliness and the de-ethnicized queer scene where any immigrant presence becomes a *foreign country*, and where a place of belonging is at once a *compelling fantasy and a lost ideal.*

References

Ahmed, S. (2004), *The Cultural Politics of Emotion* (Edinburgh: Edinburgh University Press).

Bourdieu, P. (1984), *Distinction: A Social Critique of the Judgment of Taste* (Cambridge, MA: Harvard University Press).

Boym, S. (2001), *The Future of Nostalgia* (New York: Basic Books Books).

Brah, A. (1996), *Cartographies of Diaspora: Contesting Identities* (London, New York: Routledge Books).

Butler, J. (1990), *Gender Trouble: Feminism and the Subversion of Identity* (London: Routledge Books).

Eng, D.L. and Han, S. (2003), 'A Dialogue on Racial Melancholia' in Eng and Kazanjian (eds).

Eng, D.L. and Kazanjian, D., eds (2003), *Loss: The Politics of Mourning* (Berkeley: University of California Press Books).

Fortier, A.-M. (2001), '"Coming home": Queer Migrations and Multiple Evocations of Home', *European Journal of Cultural Studies*, **4**(4), 405–424.

—— (2006), 'Community, Belonging and Intimate Ethnicity', *Modern Italy*, **1**(1), 63–77.

Freud, S. (1917/34), 'Mourning and Melancholia' in *Collected Papers* (London: Hogarth Press).

Golden, D. (2003), 'A National Cautionary Tale: Russian Women to Israel Portrayed', *Nations and Nationalism*, **9**(1), 34–51.

Kuntsman, A. (2006), *On Soldiers, Terrorists and Immigrants: Racism as Queer Homecoming.'* Paper presented at 'Out of Place: Interrogating Silences in Queerness/Raciality International Workshop', Lancaster University, Lancaster, UK, 23–25 March 2006.

Lomsky-Feder, E. and Rapoport, T. (2003), 'Juggling Models of Masculinity: Russian-Jewish Immigrants in the Israeli Army', *Sociological Inquiry*, **73**(1), 114–137.

Manalansan, M.F., IV (2003), *Global Divas: Filipino Gay Men in the Diaspora* (Durham and London: Duke University Press).

Pesmen, D. (2000), *Russia and Soul: An Exploration* (Ithaca, NY: Cornell University Press Books).

Petzen, J. (2004), 'Home or Homelike?, Turkish Queers Manage Space in Berlin', *Space and Culture*, **7**(1), 20–32.

Probyn, E. (2000), *Carnal Appetites: FoodSexIdentities* (London and New York: Routledge Books).

Ries, N. (1997), *Russian Talk: Culture and Conversation during Perestroika* (New York and London: Ithaca Press): Cornell University Press).

To USA (Tijuana, Mexico, 2006) by Elly Clarke

Chapter 9

'Abusing Our Hospitality': Inhospitableness and the Politics of Deterrence

Sarah Gibson

> Britain has a proud tradition of offering a safe haven to those genuinely fleeing persecution. We value that tradition but we cannot tolerate abuse of our asylum system. Controlled and managed migration is essential to the economic wellbeing of the UK and the health of our public services.
>
> (The Labour Party website)

Britain prides itself on its hospitality, tolerance and generosity towards refugees and asylum seekers. The above quote, taken from The Labour Party website in 2006, imagines Britain as being a hospitable and tolerant nation; the shores of Britain are imagined as providing a 'safe haven' to those 'fleeing persecution.' In contrast to Britain's hospitableness and tolerance, some of those strangers who arrive on Britain's shores are constructed as being 'abusive.' In this discourse, Britain's compassion towards strangers is used in a narrative of injury. Britain is being injured, both through its own openness to others and by those others' ungratefulness to the nation. This ungratefulness is constructed through either the stranger's 'abuse' of the system or their refusal to contribute to the 'health' and 'wellbeing' of the nation. Ironically then, in order to protect and secure British hospitality, the mobility of others needs to be 'controlled' and 'managed' by the government.

Hospitality is a dead metaphor invoked in order to reflect on 'encounter[s] with the stranger' (Rosello, 2001: 2), and Britain's compassion for those others arriving on its shores is frequently imagined using this metaphor. This chapter critically analyses this mobilization of 'hospitality' in contemporary debates on immigration and asylum in Britain and as a way of imagining British national identity. Metaphors contribute to the formation of legitimacy in political communication (Charteris-Black, 2004, 2005, 2006); hence when 'hospitality' is mobilized in current debates on immigration and asylum, it is in fact a way of justifying increasingly fortified border controls into the nation. Hospitality is invoked precisely as a way of curtailing Britain's hospitableness. The metaphor of hospitality, together with tolerance and generosity, functions as an alibi in order to protect Britain's own interests and self-image. As a headline in the *Daily Star* proclaimed in 2004, and which repeats the discourses used by the government, Britain is 'Generous to a Fault' (21 October 2004). The newspaper column argues that 'this nation has always been proud of its generosity towards the persecuted. But now we've been fleeced so much, that generosity is running out.' Britain's mythical hospitality is articulated in this rhetoric as being continuously threatened by bogus or abusive asylum seekers and economic migrants; strangers who are 'fleecing' Britain's welfare state. British hospitality is in crisis, and New Labour's response to such a crisis is to condition, limit and ration its hospitality and welcome to the other through the adoption of the principle of humane deterrence. Rather than welcoming the stranger hospitably, New Labour manages the risk of hospitality by deterring strangers from seeking refuge in Britain. It is intolerance and hostility – inhospitableness – that paradoxically endorses Britain's narcissistic pride in its self-image.

Myths of British Hospitableness

Since his election as leader of The Labour Party in 1994 and Prime Minister in 1997, Tony Blair has sought to re-brand Britain. Part of this re-branding has been the reclamation of patriotism from the Conservative Party, which had conventionally imagined Britishness in exclusive, hostile and unwelcoming terms. In contrast to the 'swamping' rhetoric employed in discourses of cultural racism by Conservative politicians such as Enoch Powell and Margaret Thatcher, New Labour instead sought to promote Britain as being a diverse, tolerant, hospitable and multicultural society. For example, Blair wrote in *New Britain* that his vision of a 'one nation' Britain was characterized as being 'tolerant, fair, enterprising, inclusive' (1996: 16). More recently in the context of British multiculturalism, Gordon Brown argued that it was precisely because 'different ethnic groups came to live together in one small island that we first made a virtue of tolerance, welcoming and including successive waves of settlers – from Saxons to Normans to Huguenots and Jews and Asians and Afro-Caribbeans' (2004). Britain's multicultural hospitable present has historically been enabled by its histories of colonial and postcolonial hospitality,[1] for example, and its openness to the 'right' kind of immigrant. Hospitality is thus a founding myth of national identity and pride in Britain.

It is considered 'a matter of national pride that persecuted people have been able to find refuge in this country' (Pirouet, 2001: 1) and it is widely believed that Britain has 'an exemplary record of offering hospitality' to those fleeing persecution (Cohen, 1994: 72). This pride in British hospitality is broadly shared across the political spectrum, with Michael Howard, the then Conservative leader, echoing the rhetoric of New Labour in 2005 by arguing that 'Britain has welcomed people from around the world with open arms. We have a proud tradition of giving refuge to those fleeing persecution. [...] Our asylum system is being abused – and with it Britain's generosity.' Three key past examples of Britain's hospitableness to strangers are frequently cited as evidence of this 'exemplary' hospitality: the Protestant Huguenots fleeing the

1 Multicultural hospitality, however, is not very hospitable as it positions the national subject in the powerful position of the host who has the capacity to give or refuse hospitality. In order to consider an ethics of hospitality, Ahmed argues that we need to consider the historical injustice of colonial hospitality, a hospitality which involved the complex 'transformation of guests into hosts' (2000: 190).

intolerance of Catholic France during the seventeenth century;[2] individuals who fled persecution in Europe because of their revolutionary politics (such as Karl Marx and Friedrich Engels); and the Jewish children from Nazi Germany who were offered sanctuary in Britain on the *Kindertransport* program from 1938 to 1940 (see Cohen, 1994; Pirouet, 2001; Cohen, 2003). These different hospitable gestures all, in very different contexts, serve to imagine and construct a British identity against its other (the intolerance of Catholic France, Europe and Nazi Germany respectively). This pride in Britain's hospitality is central for imagining a British national identity.

In this remembering of past hospitable moments there is, however, a simultaneous forgetting of past moments of inhospitableness. These myths of British hospitality conceal the fact that Britain 'has never been a haven for refugees' (Cohen, 2003: 109), and that Britain's attitude towards its others has more often than not been characterized by hostility and intolerance rather than hospitality and tolerance (Holmes, 1991). In contrast to the pride in Britain's 'exemplary' record of hospitality to migrants, refugees and asylum seekers, Louise London questions what adjective would be used to 'describe Britain's history of *not* taking in refugees: would that be proud too? Or would it be the opposite? Shamefaced? Hidden? Denied? Suppressed? Because even if it isn't proud, even if it doesn't fit the political message, this country also has a history of not taking in refugees' (2000: 18). The remembering of these past moments of hospitableness alert us to the work such myths do in present debates on nationalism, immigration and asylum in Britain. This appeal to past moments of hospitableness legitimates the present's inhospitableness, while simultaneously maintaining the promise of a future hospitableness.[3]

In the construction of a 'multiculturalist nationalism' in Britain (Fortier, 2005), the national ideal is that of a welcoming, hospitable, generous, tolerant and multicultural nation (Ahmed, 2004: 133). However, simultaneous with this pride in Britain's hospitableness, is New Labour's contribution to the moral panic over asylum in Britain. The very openness to the other, which Britain prides itself on, paradoxically raises anxieties over Britain's vulnerability to the other. While welcoming the other is the source of Britain's multiculturalism, this welcome simultaneously places the nation and its identity at risk. Despite the welcoming rhetoric of some strangers as central to the social imaginary of multicultural Britain (see Ahmed, 2000; Fortier, 2005), some strangers are simultaneously being deterred from coming to Britain. In contrast to these welcomed strangers who are now at home in Britain, there has emerged the figure of the asylum seeker which now circulates as an object of hate, fear and disgust rather than as a human subject worthy of compassion and hospitality. Articulations of British multiculturalism are played out alongside New Labour's

2 The word 'refugee' was first used in this instance to describe those French Protestants seeking refuge in England.

3 This temporal disjunction in the myths of British hospitality is noted by Pirouet (2001: 4), who comments on the paradox that on the same day, 14 June 1999, Chief Rabbi Jonathan Sacks unveiled a plaque in the House of Commons to commemorate the *Kindertransport* program, Amnesty International published a report entitled *Most Vulnerable of All: the treatment of unaccompanied refugee children in the* UK (1999). Both these events occurred the day before Labour's Asylum and Immigration Bill (1999) was read in Parliament.

policies of deterrence. Such a policy discourages potentially abusive asylum seekers from choosing to seek asylum in Britain through more restrictive, and less generous, welcomes. In making Britain a less attractive destination, this policy discourages 'asylum shoppers' from coming to Britain. The policy of deterrence promotes the positives of hospitality (a prosperous economy and a diverse culture) while defending Britain against the risk of hospitality (by excluding those others who would abuse the system).

The myth of British hospitality constructs Britain as a 'soft touch nation'. The metaphor of the 'soft touch' suggests that Britain is made 'vulnerable to abuse by its very openness to others. The soft nation is too emotional, too easily moved by the demands of others, and too easily seduced into assuming that claims for asylum [...] are narratives of truth' (Ahmed, 2004: 2). In order to protect Britain from its own compassion, such rhetoric works to legitimate increased border controls into the nation. Britain's 'softness' is thus protected through the 'hardness' of its borders.

The metaphor of the 'soft touch', of being too hospitable, is frequently connected in contemporary discourses to the 'water metaphors' of the country being flooded and swamped by asylum seekers (see Cohen, 2002: xx; Charteris-Black, 2006).[4] One example is a newspaper headline from the *Daily Express*: 'Bogus Asylum Seekers That Keep Flooding into Britain: Britain a Soft Touch on Asylum' (26 April 2001). Despite New Labour's self-conscious distancing from the cultural racism of right-wing politicians, similar rhetoric is employed in constructing the figure of the asylum seeker. For example, while introducing the Nationality, Immigration and Asylum Bill in 2002, David Blunkett, the then Home Secretary, suggested that schools were being 'swamped' by the children of asylum seekers. The verb 'swamp' in such rhetoric communicates several messages simultaneously: it evokes 'strong emotions *and* creates a myth that immigration [and asylum] is excessive *and* communicates the ideological political argument that it should be stopped' (Charteris-Black, 2006: 567).

Britain's response to the perceived 'freedom of movement' of asylum seekers has been to reassert its sovereignty and control of it borders. Asylum seekers are imagined to be hyper-mobile in their attempts and resourcefulness in reaching 'soft touch' Britain. While the figures of asylum seekers are imagined as 'flows' into the nation, the nation is imagined as a 'container' (Charteris-Black, 2006). Such metaphors arouse fear through the possibility that the boundary can be perforated, and that the country will be 'flooded' or 'swamped' by the 'flows' of asylum seekers. The perceived hyper-mobility of asylum seekers cannot be contained as the nation's borders are porous. These metaphors thus legitimate the immobilization of asylum seekers at the nation's borders, in detention centres, for example.

Such liquid metaphors of 'swamping' force connotations 'between asylum and the loss of control' (Ahmed, 2004: 46) and creates fear around the proximity

4 Metaphors of 'flow' and 'liquidity' characterize theorizations of modernity. For example, Bauman argues that 'fluidity' and 'liquidity' are 'fitting metaphors when we wish to grasp the nature of the present' (2000: 2). Urry similarly uses the metaphor of 'fluidity' to account for the 'uneven and fragmented flows of people, information, objects, money, images and risks across regions in strikingly faster and unpredictable shapes' (2000: 38).

of others. While the metaphor of 'swamping' has connotations over the loss of control over the nations' borders, the metaphor of 'hospitality' has connotations of regaining control. While Britain is swamped, and therefore passively receiving those unwelcome guests, the metaphor of hospitality raises the question of choice. *Britain chooses to be hospitable because we are generous, but we can just as easily choose to rescind that generosity*. Following from this, when asylum is framed as a political or moral obligation that is codified in international law it is perceived as a lack of control (swamping), so re-framing debates on asylum through the metaphors of generosity or hospitality enables the reassertion of the sovereignty of the nation-state. The simultaneous use of the metaphors of hospitality and the metaphors of swamping thus negotiate the tension surrounding the myth of 'British hospitality' today. It is the tension between being open (hospitable) and being abused (swamped) that is negotiated through New Labour's legislation of nationality, immigration and asylum in Britain.

Britain's Hospitable Narcissism

New Labour's pride in British multiculturalism is betrayed by the simultaneous passing through Parliament of increasingly punitive restrictions on the immigration and asylum systems. Since elected to government there has been the publication of the white paper *Fairer, Faster and Firmer* (1998) which preceded the Immigration and Asylum Act (1999), the publication of the white paper *Secure Borders, Safe Haven* (2000) which preceded the Nationality, Immigration and Asylum Bill (2002), the Immigration, Asylum and Nationality Bill (2005), and the recent five-year strategy for asylum and immigration in Britain, *Controlling Our Borders* (2005). Simultaneous with Britain's welcoming of diversity in its rhetoric of multiculturalism, is the legitimation of new border controls aimed at excluding bogus or abusive asylum seekers. The welcome to the other is thus restricted and controlled (see Tyler, 2006). The welcome offered to those genuine and deserving refugees is paradoxically enabled through the exclusion of those abusive others.

Intolerance or hostility between social groups is connected to narcissism and to what Freud identified as the 'narcissism of minor differences': 'it is precisely the minor differences in people who are otherwise alike that form the basis of feelings of strangeness and hostility between them' (1991: 272). Nationalism then is a 'kind of narcissism' (Ignatieff, 1999: 79) in its self-love for the national ideal and its aggression towards those others outside the national community. Ignatieff argues that nationalism is necessarily intolerant as it is predicated upon constructing an 'us' and a 'them.' The 'narcissism of minor differences' is thus an 'intolerance of minor differences' (1999: 86).

The narcissistic pride in Britain's hospitality and tolerance is strengthened through the distinction between those genuine, deserving, and grateful refugees, and those bogus, undeserving, abusive asylum seekers and economic migrants (Sales, 2002). The inclusion of the deserving is made possible through 'stronger boundaries of exclusion around those deemed undeserving' (Sales, 2005: 455). Maintaining these oppositions, between the deserving and the undeserving, the bogus and the genuine, the grateful and the abusive, enables Britain to 'imagine its own generosity in welcoming some others. The nation is hospitable as it allows those genuine ones to stay. And yet at the same time, it constructs some others as already hateful (as bogus) in order to define the limits of conditions of this hospitality' (Ahmed, 2004: 46). However, Derrida argues that there 'is not narcissism and non-narcissism; there are narcissisms that are more or less comprehensive, generous, open, extended. What is called non-narcissism is in general but the economy of a much more welcoming, hospitable narcissism, one that is much more open to the experience of the other as other' (1995: 199). New Labour's narcissism is therefore that of a 'hospitable narcissism' in that its ideal is located in this openness to others, even while this opening is limited and inseparable from the policies of deterrence.

These distinctions between different figures of the stranger (Ahmed, 2000) make Britain more or less hospitable, depending on whether the stranger is welcomed (and welcome-able) or not. These different welcomes are apparent in New Labour's documentation on the immigration and asylum systems. To be able of offer hospitality, British identity needs to be secure and held in place, as David Blunkett wrote in his preface to *Secure Borders, Safe Haven*:

> Confidence, security and trust make all the difference in enabling a safe haven to be offered to those coming in to the UK. To enable integration to take place, and to value the diversity it brings, we need to be secure within our sense of belonging and identity and therefore to be able to reach out and to embrace those who come to the UK (2002: 4).

This security in feeling at home locates the British subject in a position of mastery over the stranger at its door. Feeling at home enables the host to reach out and touch (and be touched by) the other by 'embracing' those welcome strangers. The 'host' thus distinguishes between those welcome and unwelcome guests. It is only those others who either genuinely 'seek to escape persecution' or 'those who wish to work and to contribute' to the nation that ultimately 'will get the welcome they deserve' (2002: 4). Britain's hospitality and welcome is restricted towards 'those who have a contribution to make to our country, offering refuge to those who have a well-founded fear of persecution and engaging those who seek citizenship so they can enjoy the full benefits of this status and understand the obligations that go with it' (Blunkett, 2002: 5). The language used distinguishes between those different strangers who are welcomed into Britain, and those who are expelled or deterred from entering. These welcome strangers are imagined as 'genuine' or 'deserving' refugees, and also those who are willing to 'contribute' to the nation. This discourse of 'cultural contribution' (Parker, 2000) figures hospitality and the welcome as a generous gift, but a gift that nevertheless demands an acknowledgement ('thank you') and which obligates the receiver in a reciprocal exchange. The gift of British hospitality is thus not (a) present as it places the guest in a 'debt of hospitality' (Chan, 2005: 21) to the British nation. Following the burning down of a detention centre by detained asylum seekers, one tabloid stated that 'this is how they thank us' (in Ahmed, 2004: 137). The welcome to Britain thus distinguishes between the grateful guest (who will pay back his/her debt to Britain through their cultural and economic contribution) and those ungrateful guests who are parasitical upon the nation state (see Gibson, 2003).

In such a discourse of Britain's excessive hospitality (the 'soft touch' nation), the figure of the abusive or bogus asylum seeker becomes associated with 'unproductive' hospitality (Derrida, 2002b: 100). Britain's hospitality is excessive in its unproductivity; nothing is given back ('contributed') to the host nation state by its ungrateful guests. New Labour's *Secure Borders, Safe Haven* concludes by asserting that 'our standards of service, our welcome to those who need our protection and who can contribute to our society and our unequivocal messages of deterrence to those who break our laws and abuse our hospitality will remain under close scrutiny' (Home Office, 2002, 8.1).

The phrases 'the abuse of hospitality' and 'abusing hospitality' have both gained currency since Labour came to power in 1997. Indeed, it is one of the 'emotive' phrases commented upon in a report analysing the representation of asylum seekers in the British press (see ICAR, 2004).[5] Charles Clarke, the Home Secretary, stated in February 2005 that 'the fairness and hospitality of the British people has been tested' by increasing numbers of refugees and asylum seekers arriving in Britain. This potential for abuse of the immigration and asylum system has resulted in the 'abuse of hospitality' (*The Guardian*, 'Labour Fuels War on Asylum', 6 February 2005). These phrases work to legitimate New Labour's radical overhaul of the immigration and asylum systems. In the Foreword to *Controlling our Borders* (2005), Tony Blair writes that British tolerance is under threat 'from those who come and live here illegally by breaking our rules and abusing our hospitality' (5). He continues to argue that it is in the nature of British people 'to be moderate and tolerant. They have, over many decades, welcomed those who desperately need a safe haven. This generosity and tolerance helps explain why race relations here have, in general, been a quiet success story. […] But this traditional tolerance is under threat' (2005: 5). Blair continues to talk of those welcomed to Britain as the nation 'benefit[s] from people from abroad who work hard and add to our prosperity' (2005: 5), while Clarke argues that 'we will continue to welcome economic migration within strict criteria. Visitors, students and migrant workers make huge contributions to the UK economy' (2005: 7). This paper reiterates that 'our welcome to those who need our protection and who can contribute to our society and our unequivocal messages of deterrence to those who break our laws and abuse our hospitality will remain under close scrutiny' (Home Office, 2005, 8.1). In order to protect Britain's capacity for hospitality in the future, there is increased surveillance and policing of both the nation's borders and those bodies who attempt to cross the borders into Britain. Such legislation thus negotiates the tension between Britain's narcissistic self-image as welcoming and hospitable, and the anxieties that accompany such openness to the other. So while such legislation attempts to define the welcome-able or abusive other in limiting the conditions of Britain's hospitableness, such legislation actually works to define Britain's narcissistic self-image as hospitable, tolerant and generous.

---

5 *The Sun*, for example, used this emotive phrase to argue that 'a sizeable proportion of those supposedly seeking a safe haven in Britain are abusing our hospitality' (29 August 2003). This hospitality is being abused, the article argues, by the criminality of asylum seekers. The alleged criminal activities by asylum seekers outlined in the article include terrorism, shoplifting, burglary, theft, and drink-driving.

The Politics of Deterrence

One of the ways British hospitality is protected has been through the government's adoption of the principle of deterrence into the immigration and asylum systems. The Acts passed through government all present an image to both the British at home and to those 'asylum shoppers' abroad that Britain is categorically not a 'soft touch,' promoting itself as an unattractive destination for those abusive asylum seekers.[6] David Blunkett suggests that while 'we should be proud that this view of the UK is held around the world,' Britain 'need[s] to send out a signal around the world that we are neither open to abuse nor a "Fortress Britain"' (2002: 5).[7] The discourse on asylum and the figure of the asylum seeker has in Britain 'rarely focuse[d] on the ethics of helping those seeking refuge' (Kushner, 2003: 257). Instead, the discourse of asylum has focused on the protection of Britain, the British way of life and its welfare system (Kundnani, 2001: 52) and on the prevention of the abuse of the system through securing the national borders and controlling those bodies who seek to cross the border in Britain.

Under New Labour, deterrence is privileged as the primary principle of Britain's asylum policy (Fekete, 2001: 24). *The Future of Multi-Ethnic Britain*, for example, suggested that the British asylum system was 'geared much more towards preventing "abuse" and discouraging arrivals, than to providing protection' (Runnymede Trust, 2001: 13). This adoption of the principle of deterrence is centred on three main areas of the asylum system in Britain: detention, dispersal and deportation (Bloch and Schuster, 2005). The deterrent asylum system is characterized by a regime that denies asylum seekers access to the welfare state in exchange for basic subsistence and shelter (Fekete, 2001: 32). So while Britain is being ostensibly hospitable,[8] it is limited in that it maintains the distinction between 'host' and 'guest' through the exclusion of asylum seekers from the host society.

Deterrence is one method that nation-states are increasingly resorting to in order to reduce 'the uninvited to their countries' and it has become 'part of the established armoury of many potential host nations' (McNamara, 1990: 123).[9] This policy of

6 The 2000 Mori Poll 'Are We an Intolerant Nation?' found that 80 per cent of British adults believed refugees came to Britain because they regarded Britain as a 'soft touch.' Research on the decision-making process by asylum seekers shows that Britain is chosen due to a variety of factions: having friends and relatives in the UK; the belief that the UK is a safe and democratic country; colonial links between their country of origin and the UK; and the English Language (see Robinson and Segrott, 2002).

7 In contrast to the 'pride' in Britain's hospitality, resistance movements are articulated through the emotion of 'shame.' The group 'Leeds No Borders' ran a guided tour of the city called 'Leeds, City of Shame' in May 2006 to increase understanding of the key local institutions involved in the destitution, dispersal, detention, and deportation of asylum seekers, while 'Welcome to Barbed Wire Britain' states on its home-page that its network 'wants to stop this shame' of detaining asylum seekers (http://www.barbedwirebritain.org.uk).

8 Hospitableness is defined as 'the giving of food, drink and accommodation to the people who are not regular members of a household' (Telfer, 1996: 83).

9 The policy of 'humane deterrence' originates from the South-East Asian response towards the arrival of Indo-Chinese immigrants during the 1980s (see McNamara, 1990).

humane deterrence denotes 'the restrictive treatment of newcomers by receiving governments in order to deter further potential asylum-seekers' (McNamara, 1990: 123). The shift towards detention as a strategy of deterrence is one that 'maximises exclusion, undermines status and rights, and emphasises short-term stay for refugees' (Joly, 1999: 336). 'Detention', as Bloch and Schuster note, has two meanings; it is both 'enclosure' within the detention centre, accommodation centre or prison, and it is 'exclusion' from the receiving host society (2005:493). 'Detention' is defined by the United National High Commission for Refugees as 'confinement within a narrowly bounded or restricted location, including prisons, closed camps, detention facilities or airport transit zones, where freedom of movement is substantially curtailed, and where the only opportunity to leave this limited area is to leave the territory' (UNHCR, 1999: Guideline 1). The use of detention is 'inherently undesirable' as the detention of asylum seekers is contrary to the norms and conventions of international refugee law (UNHCR, 1999: 1). Detention is opposed to Article 9 of the Universal Declaration of Human Rights which asserts that 'no one shall be subjected to arbitrary arrest, detention or exile.' Despite Britain's rhetoric of hospitality, in Britain 'detention is still for longer periods and with less judicial control than in [any] other European countries' (Hayter, 2004: 116).

Historically, there have been three different categories of detention in Britain: detentions from terrorist threat, wartime internments and detentions that arise from the enforcement of immigration act detainees who either violated their conditions of entry, entered illegally, or who are asylum seekers detained pending a decision on their case (see Cohen, 1994). The provision of detention in Britain was first codified in the 1971 Immigration Act, which gave immigration officials the power to detain 'any person seeking to enter or remain in the United Kingdom who is subject to immigration control' and it was 'not intended that it would routinely be used to detain asylum seekers' (Hayter, 2004: 116). As Cohen comments, this detention of asylum seekers is a contradiction in terms as social convention frequently dictates that 'a stranger with no hostile intent, or in a condition of distress, should be met with hospitality rather than the prospect of incarceration' (1994: 99). Yet, the detention of asylum seekers has increasingly become normalized within Britain (Bloch and Schuster, 2005: 491). This incarceration of asylum seekers in detention centres are 'crimes against hospitality' (Derrida, 1999: 71).

The detention centre functions as the border (the 'flood gates') into Britain. Borders are no longer simple 'lines' on the edge of a nation's territory, but are now 'detention zones' and 'filtering systems' (Balibar, 2004: 111) which are used to identify or recognize deserving or abusive asylum seekers. Despite the representation of the 'luxuriousness' of detention centres in the media's indictment of Britain's excessive hospitality and generosity, the detention centre functions as an inhospitable space in order to deter future abusive asylum seekers. This is how one asylum seeker describes Campsfield House, the largest detention centre in Britain:

> It has the aspect of a concentration camp. Its 20-foot high metal fences have mesh whose holes are slightly too small for fingers. There are 42 video cameras on the fence and inside. Periodically, as a result of protests inside and outside and attempted escapes, the fortifications are added to. Additional fences have been built parallel to the existing

> external fences. Razor wire has been put along the top of the entire length of the external fence (in Hayter, 2004: 123–124).

Asylum seekers are thus hostages to the British immigration and asylum system, and its deterrent policies of detention, dispersal and deportation.[10]

This normalization of detention is simultaneous with the demonization of the figure of the asylum seeker as an object of hate, fear and disgust. The imagining of the asylum seeker as something to be feared distances the British from those others; the dehumanization of the figure of the asylum seeker discourages ethical forms of recognition (Tyler, 2006: 195). Asylum seekers are represented as inanimate objects unworthy and undeserving of British hospitality, tolerance, generosity, compassion and empathy. It is through the 'production of the imaginary figure of the asylum-seeker as an "illegal" threat to "our" sense of national belonging that "we" learn to desire and demand "their" exclusion' (Tyler, 2006: 190–191).

Inseparable from this 'logic of deterrence' is the 'logic of suspicion' (Kundnani, 2001), which views all refugees and asylum seekers as potentially abusive. This has led to a 'culture of disbelief' in Britain, where there is the perception that refugees and asylum seekers have not travelled to Britain because of a genuine fear of persecution, but are instead merely economic migrants 'attracted to the "Honey Pot" of "Soft Touch Britain"' (Cohen, 2002: xix). Reports into the representation of asylum and asylum seekers in the UK show how the figure of the asylum seeker is negatively constructed in Britain through the emotions of fear, hate and disgust (Kaye, 1998; ICAR, 2004). Typical of the words used to construct the figure of the asylum seeker are: 'scrounger, sponger, fraudster, robbing the system, burden/ strain on resources, [...] criminal, criminal violent, arrested, jailed, guilty, mob, horde, riot, rampage, disorder, a threat, a worry, to be feared' (ICAR, 2004: 47). The *Daily Mail*, for example, writes that 'we resent the scroungers, beggars and crooks who are prepared to cross every country in Europe to reach our generous benefits system' (in Robinson and Segrott, 2002: 1). While the figure of the asylum seeker is here criminalized and parasitical, in contrast the British subject is imagined to be hospitable, tolerant and generous. It is the generosity of Britain's welfare state that needs to be protected from those who seek to abuse it. This protection of Britain's welfare and the demonization of the figure of the asylum seeker can be located in the new economic racism, where 'poverty is the new Black' (Sivanandan, 2001: 2). This 'xenoracism' is directed towards 'impoverished strangers' (Sivanandan, 2001: 2) and is symptomatic of global capitalism. This is the new 'global apartheid' which separates the rich and poor into distinct territories (Balibar, 2004: 113).

The infinite swamping of Britain by such abusive asylum seekers injures Britain. The 'culture of disbelief', whereby any body crossing the border into Britain could potentially be an abusive asylum seeker, justifies 'the repetition of violence against the bodies of others in the name of protecting the nation' (Ahmed, 2004: 47). In constructing the figure of the abusive asylum seeker, Britain actually seeks to redefine its own national character as such 'processes of exclusion and rejection uncover and

10 'Hostage' is etymologically linked to the series host, guest and ghost (Derrida, 1993: 160).

reveal and become constitutive of the national identity itself' (Cohen, 1994: 198). In such a limited model of hospitableness, the figure of the abusive asylum seeker is constructed in order to support Britain's pride in its ideal.

The Promise of Hospitality

In the rhetoric around asylum and asylum seekers in Britain today, Britain is paradoxically either too hospitable or not hospitable enough. Debate seems to be focused on those who are proud of the country's hospitable past (a past which legitimates border controls in the present) or ashamed of our country's inhospitable present. This aporia of hospitality is in fact the condition of hospitality for Jacques Derrida. Derrida's recent writings have focused on hospitality, generosity and tolerance as the figures of the impossible. However, for Derrida there is always a tension between the limits of a conditional hospitality and an infinite unconditional hospitality. To talk of a 'British' hospitality then, is somewhat misleading for it is in fact impossible for a nation-state to be properly hospitable.

In his writings on hospitality, Derrida charts how 'hospitality' shares the same etymological root as 'hostility' (1997a: 3). Hospitality then is always parasitized by hostility, its apparent opposite. Derrida's neologism of 'hostipitality' discloses this impossibility of an absolute hospitality (2002a). Derived from the Latin *hostis* (the 'hostile' stranger) and *potis* (to have power), the social relations constructed through gestures of hospitableness are thus implicated in power relations, where it is the host who has both the power and the property to give to the stranger, but crucially while remaining in control and ownership. Hospitableness, tolerance, generosity and compassion are therefore all distancing gestures that maintain the opposition of self and other.

This power relation of hospitality constitutes the aporia of hospitality for Derrida, as 'it does not seem to me that I am able to open up or offer hospitality, however generous [...] without reaffirming this is mine' (1997a: 14). In welcoming the stranger to the home, the host asserts the position of mastery through his/her identification with the home. It is through welcoming the other that the host is able to 'insinuate that one is at home here, that one knows what it means to be at home, and that at home one receives, or offers hospitality, thus appropriating for oneself a place to *welcome* the other' (Derrida, 1999: 15–16). The government (the hosts) have welcomed the genuine and deserving asylum seeker through asserting their ownership and control of the nation (home). But this welcome is strictly limited and conditional upon their guest's gratefulness and conformity to British values. This hospitality is offered on the condition that 'the other follow our rules, our way of life, even our language, our culture, our political system, and so on' (Derrida, 2003: 128). Britain's hospitality is necessarily conditional in that it consists of 'welcoming particular guests and [...] as a result, not others' (Naas, 2003: 164). New Labour's policies have attempted to minimize the risk of abusive asylum seekers through their adaptation of a policy of deterrence. This is a conditional hospitality, in contrast to the absolute hospitality which consists of 'leaving one's house open to the unforeseeable arrival' (Derrida and Roudinesco, 2004: 59). This is the risk implicit in any gesture of hospitality, which is controlled and managed with respect to the figure of the asylum seeker. As objects of fear, disgust and danger, 'asylum seekers defy calculations of risk as [...] they are "unknowable", "ungovernable" and thereby "dangerous"' (Malloch and Stanley, 2005: 54). To be properly hospitable, the law of hospitality invites us 'to welcome anyone [...] without checking at the border' (Derrida, 1997a: 8). This law of absolute hospitality, of the unconditional welcome, however, is distinct from the laws of a political and conditional hospitality. This is the impossibility of a national hospitality for Derrida, as the nation-state is constituted through its control and securing of its borders; no nation-state finds it acceptable to abandon its border controls and immigration controls (2005: 6). This policing and securing of the nation's borders simultaneously poses and limits the question of hospitality, for 'a nation can not not suspend [...] this principle of absolute hospitality' (Derrida, 2005: 6). British national sovereignty 'can only be exercised by filtering, choosing, and thus by excluding and doing violence' (Derrida, 2000b: 55).

While New Labour celebrates the nation through the metaphors of hospitality and tolerance, both have concealed power relations which work to fix the other in place. Tolerance secures the British subject in a powerful position of mastery or sovereignty. The limit of a nation's hospitality is its tolerance (Derrida, 2003: 127–128). Tolerance limits the welcome, working precisely to 'retain power and maintain control over the limits of my "home"' (Derrida, 2003: 127–128). Tolerance then is 'a conditional, circumspect, careful hospitality' (Derrida, 2003: 128). New Labour's endorsement of tolerance is thus ambiguous as it maintains the binary opposition of self/other, host/guest, British/asylum seeker (Lewis, 2005; Wemyss, 2006). In advocating Britain's tolerance, there is still the capacity for Britain's intolerance. The apparent opposition of 'tolerant nationalism' and 'intolerant nationalism' is thus false, as both employ similar 'practices of spatial power' (Hage, 2000: 91). In this respect, Britain imagines itself through its encounters with strangers, whether their proximity is threatening or welcomed (Ahmed, 2000: 100). 'Welcoming' the other then, is premised on the same mastery of 'expelling' the other, as both encounters are 'a means by which the "we" asserts itself' (Ahmed, 2000: 190). The hospitality of a nation is therefore dependent upon its thresholds of tolerance, and the social tolerance of asylum seekers has today 'reached its limits' (Malloch and Stanley, 2005: 56).

Asylum is based on an ethical relational to the law's abject classes (such as asylum seekers). In this respect asylum is excessive hospitality which necessarily needs to be limited by the laws of the nation state. Absolute hospitality requires the gift or generosity of the state even as the ethical notion of this welcome goes beyond the political and juridical confines of the state. It is time, as Derrida asserts, to 'call out for another international law, another border politics, another humanitarian politics, indeed a humanitarian commitment that *effectively* operates beyond the interests of nation-states' (1999: 101).

While current laws on immigration and asylum in Britain limit hospitality, such laws are 'infinitely perfectible' (Derrida, 2002b: 104). An unconditional welcome is always to come (Derrida, 2000b: 55). This unconditional hospitality or welcome is 'impossible to live' and cannot have 'legal or political status' in a nation's laws, 'but without at least the thought of this pure and unconditional hospitality, of hospitality itself, we would have no concept of hospitality in general and would not even be able to determine any rules for conditional hospitality. [...] Unconditional hospitality, which is neither juridical nor political, is nonetheless the condition of the political and the juridical' (Derrida, 2003: 129). The law of hospitality is always pervertible and perfectible, as it is 'as through the categorical imperative of hospitality commanded that we transgress all the laws [...] of hospitality' (Derrida, 2000b: 75).[11] Hospitality is always to come, as 'not only will it remain indefinitely perfectible, hence always insufficient and future, but, belonging to the time of the promise, it never exists, it is never present, it remains the theme of a non-presentable concept' (Derrida, 1997b: 306).

Despite this failure of any nation-state being able to offer an unconditional hospitality to the other, this does not necessarily mean that the imperatives of hospitality, tolerance or generosity should be abandoned. As Ahmed argues, the emotions of pride and shame are intimately connected (2004). While New Labour has invoked pride in past histories of hospitality in order to legitimate present inhospitableness, and campaigners for the rights of asylum seekers have invoked discourses of shame in Britain's current 'crimes against hospitality' which are endured by the 'hostages of our times' (Derrida, 1999: 71), both these discourses of pride and shame in the national ideal maintain a belief in the promise of hospitality. While the national 'politics of shame' exposes the failure of the nation in living up to its hospitable ideal, 'at the same time *it involves a narrative of recovery as the re-covering of the nation*' (Ahmed, 2004: 112). There is then hope in recovering this ideal of hospitality in the future.

Despite British hospitality being paradoxically limited in the name of British hospitality, this conditioning and rationing bears the trace of the unconditional and pure hospitality which Derrida discusses. While this unconditional hospitality threatens the sovereignty of the nation-state, any act of hospitality bears the trace of this imperative. While the national laws of hospitality, its immigration and asylum policies, are by their very nature unwelcoming to the unknown other, these laws are still imagined with reference to the law of hospitality. This is the promise of British hospitality.

References

Ahmed, S. (2000), *Strange Encounters* (London: Routledge Books).

—— (2004), *The Cultural Politics of Emotion* (Edinburgh: Edinburgh University Press).

Amnesty International (1999), *The most Vulnerable of All: The Treatment of Unaccompanied Refugee Children in the UK* (London: Amnesty International).

Balibar, E. (2004), *We, the People of Europe?* (Oxford: Princeton University Press Books).

11 New Labour's commitment to protecting British hospitality (through adopting the principle of deterrence, for example) is thus a perversion of the law of hospitality. Derrida argues that in perverting this law 'one can become virtually xenophobic in order to protect or claim to protect one's own hospitality, the own home that makes possible one's own hospitality' (2000b: 53).

Blair, T. (1996), *New Britain* (London: Fourth Estate Books).
—— (2005), 'Foreword' in *Controlling Our Borders*. The Home Office (ed.) (London: HMSO Books).
Bloch, A. and Schuster, L. (2005), 'At the Extremes of Exclusion: Deportation, Detention, and Dispersal', *Ethnic and Racial Studies*, **28**(3), 491–512. [DOI: 10.1080/0141987042000337858]
Blunkett, D. (2002), 'Foreword', in The Home Office, *Secure Borders, Safe Haven* (London: HMSO).
Brown, G. (2004), The British Council Annual Lecture, accessed online at http://www.hm-treasury.gov.uk/newsroom_and_speeches/press/2004/ press_63_04.cfm (accessed:10.11.06).
Buaman, Z. (2000), *Liquid Modernity* (London: Polity Press).
Chan, W. (2005), 'A Gift of a Pagoda, the Presence of a Prominent Citizen, and the Possibilities of Hospitality', *Environment and Planning D: Society and Space*, **23**(1), 11–28.
Charteris-Black, J. (2004), *Corpus Approaches to Critical Metaphor Analysis* (London: Palgrave Macmillan Books).
—— (2005), *Politicians & Rhetoric* (London: Palgrave Macmillan Books).
—— (2006), 'Britain as a Container: Immigration Metaphors in the 2005 Election Campaign', *Discourse and Society*, **17**(5), 563–582. [DOI: 10.1177/0957926506066345]
Clarke, C. (2005), 'Foreword' in *Controlling our Borders*. The Home Office (ed.) (London: HMSO Books).
Cohen, R. (1994), *Frontiers of Identity* (London: Longman).
—— (2002), *Folk Devils and Moral Panics*, 3rd edn (London: Routledge Books).
—— (2003), *No One is Illegal* (Stoke-on-Trent: Trentham Books Ltd.).
Derrida, J. (1993), *Aporias*. Dutoit, T. (trans.) (Stanford: Stanford University Press Books).
_____. (1995), *Points: Interviews, 1974–1994*, ed. by E. Weber, trans. by P. Kamuf (Stanford, CA: Stanford University Press Books).
—— (1997a), 'Perhaps or Maybe' in Dronsfield and Midgley (eds).
—— (1997b), *Politics of Friendship*. Collins, G. (trans.) (London: Verso Books).
—— (1999), *Adieu to Emmanuel Levinas*. Brault, P. and Naas, M. (trans.) (Stanford, CA: Stanford University Press Books).
—— (2000), 'Hostipitality', *Angelaki*, **5**(3), 3–18. [DOI: 10.1080/09697250020034706]
—— (2000b), *Of Hospitality*. Bowlby, R. (trans.) (Stanford, CA: Stanford University Press Books).
—— (2002a), *Acts of Religion*. Anidjar, G. (ed.) (London: Routledge Books).
—— (2002b), *Negotiations: Interventions and Interviews 1971-2001*. Rottenberg, E. (ed. and trans.) (Stanford: Stanford University Press Books).
—— (2003), Interviewed in G. Borradori, *Philosophy in a Time of Terror* (Chicago, IL: University of Chicago Press).
—— (2005), 'The Principle of Hospitality' *Parallax* **11**(1), 6–9.
Derrida, J. and Roudinesco, E. (2004), *For what Tomorrow… A Dialogue* (Stanford, CA: Stanford University Press Books).

Dronsfield, J. and Midgley, N., eds (1997), 'Responsibilities of Deconstruction', *PLI – Warwick Journal of Philosophy*, **6**, 1–18.

Fekete, L. (2001), 'The Emergence of Xeno-Racism', *Race and Class*, **43**(2), 23–40. [DOI: 10.1177/0306396801432003]

Fortier, A. (2005), 'Pride Politics and Multiculturalist Citizenship', *Ethnic and Racial Studies*, **28**(3), 559–578. [DOI: 10.1080/0141987042000337885]

Freud, S. (1991), *On Sexuality*, Penguin Freud Library Volume 7, trans. by J. Strachey (Harmondsworth: Penguin).

Gibson, S. (2003), 'Accommodating Strangers: British Hospitality and the Asylum Hotel Debate', *Journal for Cultural Research*, 7(4), 367–386. [DOI: 10.1080/1479758032000165039]

—— (2006), '"The Hotel Business is about Strangers": Border Politics and Hospitable Spaces in Stephen Frears' *Dirty Pretty Things*', *Third Text*, **20**(6), 691–699. [DOI: 10.1080/09528820601069631]

Hage, G. (2000), *White Nation* (London: Routledge Books).

Hayter, T. (2004), *Open Borders*, 2nd edn (London: Pluto Publishing).

Hesse, B., ed. (2000), *Un/Settled Multiculturalisms* (London: Zed Books).

Holmes, C. (1991), *A Tolerant Country* (London: Faber & Faber Books).

Home Office (2002), *Secure Borders, Safe Haven: Integration with Diversity in Modern Britain* (London: HMSO Books).

—— (2005), *Controlling our Borders: Making Migration Work for Britain* (London: HMSO Books).

Howard, M. (2005), 'Firm but Fair Immigration Controls', Speech at Conservative Campaign Headquarters (24 January 2005). Available at http://www.conservatives.com.

Ignatieff, M. (1999), 'Nationalism and Toleration' in Mendus (eds).

ICAR (Information Centre about Asylum and Refugees in the UK) (2004), *Media image, community impact: Assessing the impact of media and political images of refugees and asylum seekers on community relations in London.* Available at http://www.london.gov.uk/mayor/refugees/docs/mici_full_report.pdf> (accessed: 20.08.07).

Joly, D. (1999) 'A New Asylum Regime in Europe' in Nicholson and Twomey (eds).

Kaye, R. (1998) 'Redefining the Refugee: the UK Media Portrayal of Asylum Seekers' in Koser and Lutz (eds).

Koser, K. and Lutz, H., eds (1998), *The New Migration in Europe* (Basingstoke: Macmillan Publishing).

Kundnani, A. (2001) 'In a Foreign Land: The New Popular Racism', *Race and Class*, **43**(2), 41–60. [DOI: 10.1177/0306396801432004]

Kushner, T. (2003), 'Meaning Nothing but Good: Ethics, History and Asylum-Seeker Phobia in Britain', *Patterns of Prejudice*, **37**(3), 257–276. [DOI: 10.1080/0031322030759З]

Lewis, G. (2005), 'Welcome to the Margins: Diversity, Tolerance, and the Policies of Exclusion', *Ethnic and Racial Studies*, **278**(3), 536–558. [DOI: 10.1080/0141987042000337876]

Loescher, G. and Monahan, L., eds (1990), *Refugees and International Relations* (Oxford: Clarendon Press Books).

London, L. (2000), 'Whitehall and the Refugees: The 1930s and the 1990s', *Patterns of Prejudice*, **34**(3), 17–26.

Malloch, M. and Stanley, E. (2005), 'The Detention of Asylum Seekers in the UK', *Punishment and Society*, **7**(1), 53–71. [DOI: 10.1177/1462474505048133]

McGhee, D. (2005), *Intolerant Britain?* (Maidenhead: Open University Press Books).

McNamara, D. (1990) 'The Origins and Effects of "Humane Deterrence" Policies in South-East Asia' in Loescher and Monahan (eds).

Mendus, S., ed. (1999), *The Politics of Toleration* (Edinburgh: Edinburgh University Press).

MORI (2000), *Britain Today – Are We an Intolerant Nation?*, available at http://www.ipsos-mori.com/polls/2000/rd-july.shtml (accessed: 10.11.06)

Naas, M. (2003), *Taking on the Tradition* (Stanford: Stanford University Press Books).

Nicholson, F. and Twomey, P., eds (1999), *Refugee Rights and Realities* (Cambridge: Cambridge University Press).

Parker, D. (2000), 'The Chinese Takeaway and the Diasporic Habitus' in Hesse (eds).

Pirouet, L. (2001), *Whatever Happened to Asylum in Britain?* (Oxford: Berghahn Books).

Robinson, V. and Segrott, J. (2002), 'Understanding the Decision-Making of Asylum Seekers', Home Office Research Study 243. Accessed online at: http://www.homeoffice.gov.uk/rds/pdfs2/hors243.pdf (accessed: 10.11.06).

Rosello, M. (2001), *Postcolonial Hospitality* (Stanford: Stanford University Press Books).

Runnymede Trust (2000), *The Future of Multi-Ethnic Britain: The Parekh Report* (London: Profile Books Books).

Sales, R. (2002), 'The Deserving and the Undeserving?, Refugees, Asylum Seekers and Welfare in Britain', *Critical Social Policy*, **22**(3), 456–478. [DOI: 10.1177/0261018302022003293]

—— (2005), 'Secure Borders, Safe Haven: A Contradiction in Terms?', *Ethnic and Racial Studies*, **28**(3), 445–462. [DOI: 10.1080/0141987042000337830]

Sivanandan (2001), 'Poverty is the New Black', *Race and Class*, **43**(2), 1–5. [DOI: 10.1177/0306396801432001]

Telfer, E. (1996), *Food for Thought: Philosophy and Food* (London: Routledge Books).

Tyler, I. (2006), '"Welcome to Britain": The Cultural Politics of Asylum', *European Journal of Cultural Studies*, **9**(2), 185–202. [DOI: 10.1177/1367549406063163]

UNHCR (1999), UNHCR Revised Guidelines on Applicable Criteria and Standards Relating to the Detention of Asylum Seekers. Available at http://www.unhcr.org (accessed: 10.11.06).

Urry, J. (2000), *Sociology Beyond Societies* (London: Routledge Books).

Wemyss, G. (2006) 'The Power to Tolerate: Contests over Britishness and Belonging in East London', *Patterns of Prejudice*, **40**(3), 215–236. [DOI: 10.1080/00313220600769406]

Vietnam to Laos (Lao Bao, 2004) by Elly Clarke

Chapter 10

Hospitality and the Limitations of the National

Karima Laachir

> It is as though hospitality were the impossible: as though the law of hospitality defined this very impossibility, as if it were only possible to transgress, as though the law of absolute, unconditional, hyperbolical hospitality, as though the categorical imperative of hospitality commanded that we transgress all the laws (in the plural) of hospitality namely the conditions, the norms, the rights and the duties that are imposed on hosts and hostesses, on the men and women who receive it. And vice versa, it is as though the laws (plural) of hospitality, in marking limits, power, rights, and duties, consisted in challenging and transgressing the law of hospitality, the one that would command that the "new arrival" be offered an unconditional welcome (Derrida, 2000b: 75).

Hospitality as an ancient tradition with ethical imperatives has become politicized in Europe and the New World in the last two decades. Strict hospitality laws have been issued to 'protect' rich states from any form of visitation from poor countries since they are perceived as potential economic immigrants that may threaten the financial, social and political stability of the host countries. The mobility of non-European nationals such as Africans, Middle Easterners and Asians are perceived with mistrust as potentially undesirable 'guests' in the rich North, whereas the movement of capital, goods and 'tourists' is unlimited towards countries of the South. The imbalance of this relationship calls for an analysis of the relationship between hospitality and capital and property, especially the national one.

The European popular imagination has been haunted by images of Europe being inundated by foreigners – economic and political refugees – perceived as 'welfare-scroungers', 'job-snatchers' and 'threats to security'.[1] Some politicians have started to foment these fears to pick up extra votes, especially extreme right-wing movements, which have been gaining ground in local and parliamentary elections.[2] The increasing popularity of leaders of far right parties, who all publicly voice their xenophobia and racism against those perceived as foreigners, are alarming examples of the return of exclusionist popular nationalism and fascism to haunt postcolonial

1 *The Economist*, May 2000: 25–26, 31.

2 With their xenophobic and ethno-nationalist politics, extreme right-wing parties are represented in a number of European parliaments, such as the Austrian, Belgian, Danish, Swiss, Norwegian and Italian. They are also represented in local and regional councils in France, Germany and most recently in the UK (see Rydgren, 2005: 414).

Europe. 'Immigration' demands and those of ethnic minorities, especially religious demands, have become contentious issues in Europe. Hospitality has become more difficult since the 9/11 attacks and the subsequent 'war on terror' led by the American Government. The terrorist bombings in Madrid (March 2004) and London (July 2005) have been interpreted by some as a conflict between contending civilizations, Western and Islamic. The lives of diasporic Muslims and of immigrants in Europe and the United States have become subject to constant surveillance and are the subject of various regulations that aim to keep Muslim Fundamentalist networks under control. However, the lives of ordinary European Muslims have been deeply affected by these changes and, as a result, their loyalty, together with their European citizenship and strong cultural affiliation to Europe as their homeland, have been brought into question. They are now viewed with distrust and caution.

Hospitality is important, therefore, as an analytical concept since it opens up the debates of welcoming 'otherness' beyond issues of the reception of immigrants by their 'host' countries, towards more important problems of living together with people of 'different' cultural, religious and social affiliations. More than ever before, the world is a melting pot of different cultures and thus we are confronted with the theme of how to survive with the 'other', or those perceived as others, without seeing them as a threat or danger. The problem of xenophobia and racism (which is not limited to Europe) in the last decades after the horrors of colonialism and fascism raises a crucial question about the relationships between communities of different 'race', religion and culture. The 'us' and 'them' differentiation – camouflaged in various discourses: 'ethnic' (a soft word for 'racial'), 'religious', but mainly cultural terms – is marked by a strong degree of xenophobia, fear and racism. Technological and communicative revolutions, economic and political upheavals, such as de-industrialization, unemployment, poverty and the mass displacement of populations are all factors that have 'once again invited many to find in populist ultranationalism, racism, and authoritarianism, reassurance and a variety of certainty that can answer radical doubts and anxieties over self-hood, being, and belonging' (Gilroy, 2000: 155).

This chapter engages with Derrida's critique of the concept of hospitality in Western philosophy and culture, which he defines as being a conditional hospitality, a hospitality of invitation and not visitation. You invite someone to your country, to your house and you set the rules for that invitation. In that sense, your welcome of the other remains limited by law and jurisdiction. This type of hospitality, according to Derrida, does not interrupt the mastery of the host over his/her home or national space, quite the opposite; it is a reassertion of that mastery. Unconditional hospitality, on the other hand, is about allowing the self to be interrupted or questioned by the welcome of the other, that is, to welcome the other without setting restrictions or limitations. My question is how can we use Derrida's idea of the intervention of unconditional hospitality or ethics in the making of conditional hospitality or politics at a time when hospitality is marked by closure and fear, especially in France, his 'home' country? I examine the way hospitality is marked by an 'inclusive exclusion' of Europe's postcolonial settlers, who are still perceived as aliens with no links to their host country and who are viewed as a threat to the uniformity and integrity of the nation. I argue that the attempt to fix the social, economic and cultural mobility of these diverse postcolonial diasporic communities is a manifestation of the perpetuation of colonial culture that still preserves the same power structures that existed in the colonies.

Hospitality and National Interests

Immanuel Kant's ideas on cosmopolitanism and world citizenship have been important in framing contemporary debates on hospitality. Kant (1957: 21) envisages universal hospitality as a condition of perpetual peace and world citizenship. The globe is a restricted sphere in which we are bound to live in each other's company or to move from one place to another, and this can only be guaranteed through the right to reciprocal hospitality. It is only through hospitality that humanity can gradually be brought closer to a constitution establishing world citizenship and thus perpetual peace. Kant dismisses hospitality as philanthropy and insists on its being a right or a 'natural law'. He criticizes colonialism and the 'inhospitable actions of the civilised', referring to the commercial states of Europe which advanced their economic exploitation of territories seen as 'virgin lands' or lands without inhabitants (Africa, the Cape and America), and also for the way they used their military superiority to subdue the local populations of the newly discovered realms (Kant, 1974: 20–21). However, Kant still links hospitality with commerce in 'Of the Guarantee for Perpetual Peace', in which he argues that though 'nature wisely separates nations', it is trade and commerce that subdue the spirit of war (which is for him the state of nature) (1957: 32). It is through the power of wealth that states find themselves forced to pursue 'noble causes' and thus to search for peace, but Kant does not address the problem of how peace may be decided differently between those who have wealth and influence and those who have not. Thus, his cosmopolitanism is exclusive to certain powerful states that pass the law on the rest of humanity. Even though Kant's ideas of cosmopolitanism, universal hospitality and common right to the surface of the earth shared by all human beings have had a strong appeal in contemporary debates on democracy and citizenship, his 'racial theories' sit uncomfortably and embarrassingly with his claims to metropolitanism marked by exclusiveness. His democratic aspirations could not contain the black 'race', as his raciological ideas about the inferiority of the 'Negro' and his warning against the dangers of racial mixing contradict his cosmopolitanism.[3] Kant's universal rights of citizenship and hospitality are exclusive to those who are recognized as having a 'universal self'. Thus, those who are not recognized as having the particular cultural and corporeal attributes that announce their possession of a universal self were exempt from the moral and civic rules of conduct. In other words, if their 'race', religion, colour or nationality deprived them of access to human universal selfhood, they would be in great danger (Gilroy, 2000: 61). Humanity, therefore, was restricted to specific territorial boundaries of racialized nation-states. The history of this Western exclusionary humanism is bluntly clear in the history of slavery and

3 Kant has developed these ideas in a number of his works, especially *Anthropology From a Pragmatic Point of View* (1798) and his 1775 essay 'On the Different Races of Man' (see Gilroy, 2000: 59 for an extensive discussion of Kant's raciological thinking).

colonialism. Though Kant shows resistance to the project of colonialism, his ideas about the black 'race' and its inferiority weakens his democratic hopes and dreams. It is this exclusionary aspect of hospitality that must be questioned, especially the exclusion and marginalization of the descendants of post-war migrant settlers from society, despite their European citizenship, on the basis of those 'cultural', 'ethnic', 'religious' and social affiliations that are deemed incompatible with 'European values'.

Kant's universal hospitality as a condition for world peace does not leave any space for any form of ethical consideration as it is solely based on the 'legal' or the juridical. In light of this, Derrida (2001: 22) accuses Kant of restricting hospitality to state sovereignty, as he defines it as a law: 'Hospitality signifies here the *public nature* (publicité) of public space, as it is always the case for the juridical in the Kantian sense; hospitality is dependent on and is controlled by the law and the state police'. Kant limits universal hospitality to a number of juridical and political conditions (it is first limited only to citizens of states, it is only temporary, and so on) which, though institutional, are based on a common 'natural right' of the possession of the surface of the earth. Unlike Kant, Emmanuel Levinas introduces the disjunction between the host and the guest, the host becoming the guest of the guest in his/her own home as the home of the other, that is, to be welcomed by the face of the other that one intends to welcome. In *Totalité et Infini*, Levinas (1961) criticizes the 'tyranny of the state' when hospitality becomes part of the state or becomes political because even though this becoming political is a response to the call of the third and a response to an 'aspiration', it still deforms the I and the other and thus introduces 'tyrannical violence'. Politics, therefore, should not be left on its own, because in Levinas's words 'it judges them [the I and the other who have given rise to it] according to universal rules, and thus as [being] *in absentia*'.[4] In other words, the political renders the face invisible at the moment of bringing it into the space of public phenomenality.

In *Adieu To Emmanuel Levinas* (1999a: 21), Derrida reflects on Levinas' *Totalité et Infini*, which he perceives as 'an immense treatise of hospitality'. In this treatise, Levinas insists that the face that must be welcomed, must not be reduced to 'thematization' (*thématisation*) or description, and neither must hospitality. The face refers to the infinite alterity of the other who is free from any theme and who cannot be described. In other words, the other cannot be possessed or mastered. Hospitality, therefore, is opposed to *thematization* because it is the welcoming of the other who cannot be calculated or known, that is, the other is infinite and 'withdraws from the theme' (Derrida, 1999a: 23). Welcoming or receiving in the Levinasian sense implies the act of receiving as an ethical relation. Thus, the welcome to come presupposes 'recollection' (le recueillement) or the 'the intimacy of the at-home-with-oneself'. He claims that the 'at-home-with-oneself' does not mean to close oneself off, but rather is a 'desire' towards the transcendence of the other (Derrida, 1999a: 92). Therefore, Levinas recognizes that there can be no welcome of the other or hospitality without

4 Cited in Derrida (1999a: 98). Originally, in Levinas (1961: 276) 'elle les juge [le moi et l'autre qui l'ont suscitée] selon les règles universelles et, par là même, comme par contumace.'

this radical alterity which in turn presupposes 'infinite separation'. Thus, 'the at-home-with-oneself would thus no longer be a sort of nature or rootedness but a response to a wandering, the phenomenon of wandering it brings to a halt' (Derrida, 1999a: 92). Levinas suggests a theory of respecting the other instead of 'mastering' him/her; that is, a theory of desire that bases itself on infinite separation instead of negation and assimilation.[5] Levinas attempts to change the conventional tradition of the relation to alterity as an appropriation of the same in its totality to a different mode of relation based on respect of the infinity and heterogeneity of the other.

Hospitality in the Levinasian sense also presupposes feminine alterity.[6] Hospitality comes before or precedes property and thus its law dictates that the host who welcomes the invited or received guest is in truth a guest received in his own home. It is this absolute precedence of the welcome where the master of the house is already a received *hôte* (host) or a guest in his own home, that would be called 'feminine alterity' ('l'altérité feminine'). The pervertible or perverting nature of the law of hospitality implies that absolute hospitality should break with hospitality as a pact or a right or duty, as the former means the welcoming not only of the foreigner but of the absolute, unknown other. What is needed today in comparing Kant and Levinas, and with regard to the right of refuge in a world of millions of displaced people, Derrida argues (1999a: 101), is to 'call out for another international law, another border politics, another humanitarian politics, indeed a humanitarian commitment that effectively operates beyond the interests of the Nation-States.'

France, Derrida's 'home' country, has tended since the Revolution to distinguish itself from other European countries by portraying itself as a country of asylum seekers. The motives behind this opening up of policy have never been purely 'ethical' as 'the law of hospitality', 'a moral law', or 'the law of the land (séjour)-(ethos)' (Derrida, 2001: 10). An economic reason has been behind this policy, as the decrease in the birth rate since the middle of the eighteenth century has been behind France's 'liberal' policies in matters of immigration, especially when it is in desperate need of workers. This was certainly clear in the case of the 1960 migration when migrant workers mostly from North Africa were needed. Moreover, the right to asylum in France has only recently become juridical as it was only in 1954 that the definition of the asylum seeker (which was limited to those persecuted because of their defence of liberty) was broadened to include those whose lives were threatened by reason of race, religion or political opinion. The supposed ethical principles of the Enlightenment have been strictly controlled by juridical institutions as the (demographic and economic) interests of the nation states have restricted them (Derrida, 2001: 10–12. This absence of ethical considerations is analysed in a

5 For a detailed analysis, see Young (1990: 13–17).

6 Taking into consideration the criticism directed at Levinas's sexist language, Derrida (1999a: 44) defends Levinas's choice of 'feminine being' which is different from 'the fact of empirical woman': 'the welcome, the anarchic origin of ethics belongs to the "dimension of femininity" and not to the empirical presence of a human being of the "feminine sex".'

number of Derrida's works on hospitality, especially the conditional welcome offered to postcolonial migrants and their descendants in most European countries.[7]

Drawing on Benveniste's etymological analysis of the term hospitality and its origins in Indo-European languages, Derrida (2000b: 145–151) argues that *hostis* reveals a strange crossing between enemy and host. This is due to the troubling analogy in their common origin between *hostis* as host and *hostis* as enemy and thus between hospitality and hostility or what Derrida calls *hostipitality*: hospitality carrying within it the danger of hostility. If hospitality as a concept carries within it its own contradiction – hostility – and if 'hospitality is a self-contradictory concept and experience which can only self-destruct (put otherwise, produce itself as impossible, and only be possible on the condition of its possibility) or protect itself from itself, auto-immunise itself in some way, which is to say, deconstruct itself – precisely – in being put into practice' (Derrida, 2000a: 5), how can we talk about a politics of hospitality? If hospitality carries the danger of hostility – 'hostipitality' as Derrida calls it, while he tries to keep us alive to the dangers of fixing the threshold of hospitality and its meanings, a fixation that results in hostility or violence at the moment of welcoming the guest – how can the right to universal hospitality be guaranteed? Can one offer hospitality and remain the master of the house, the master in one's household, one's city, one's nation and one's state? Can we speak of an ethics of hospitality? Or in other words, how can we regulate in a specific juridical and political set of laws the infinite and unconditional hospitality? How can we implement this ethics as a regulatory power behind politics or laws?[8]

According to Derrida, hospitality in the 'Western' tradition is marked by the paternal and the phallogocentric, or by the logic of the master/host, nation, the door or the threshold. His critique calls into question the limitations of this specifically 'European' history of hospitality and suggests a future beyond this history, and thus a hospitality beyond the logic of 'paternity' (and its extension to the nation) or the *logos*. This does not mean that nation states should open their borders unconditionally to any 'new' comer or that they should go beyond their national interests to 'welcome' the other. In fact, Derrida's critique is a call to resist the tyranny of the state and its law making while opening up democratic institutions beyond a certain patriotic reductionism. That is what Derrida calls his 'New International', a rebellion against patriotism: 'compatriots of every country, translator-poets rebel against patriotism' (1997a: 57). Hospitality lives on the paradox of presupposing a nation, a home, a door for it to happen but once one establishes a threshold, a door or a nation, hospitality ceases to happen and becomes hostility (Derrida, 2001: 6). Therefore, hospitality is marked by a double bind and its impossibility is the condition of its possibility. It

---

7 Derrida expressed his grief when he first heard the expression *délit de l'hospitalité* (crime of hospitality) which is a law in France that allows the prosecution of those who offer their hospitality to the ones deemed as 'illegal' or *sans papiers*. What would become of a country and a culture where hospitality is seen as a crime in the eyes of the law? (Derrida, 1997b: 3–8).

8 The 'certain places' that hospitality and its quasi-synonym 'welcome' point us to, Derrida argues, are the places of the 'birth of the question', places the ethics, politics and the law (Derrida, 1997b: 45).

stays on the threshold that keeps it alive and open to new-comers. The distinction introduced in Derrida's works between, on the one hand, unconditional hospitality or 'absolute desire for hospitality' and on the other, conditional hospitality or the rights and duties that condition hospitality ('a law, a conditional ethics, a politics') is not a distinction that 'paralyses' hospitality. In fact, it aims at directing our attention to find an 'intermediate schema' between the two, 'a radical heterogeneity, but also indissociability' in the sense of calling for the other or prescribing the other. To keep alive the aporia between ethics (the law of hospitality) and politics (the laws of hospitality) is to keep political laws and regulations open to new changes and circumstances and to keep alive the fact that hospitality is always inhabited by hostility. It is the question of intervening in the conditional hospitality in the name of the unconditional, an intervention that, though surrounded by contradictions and aporias, recognizes the need of 'perverting' the laws for the sake of 'perfecting' them. Derrida stresses the aporetic relationship between the unconditional hospitality or ethics, which starts with risks, and the conditional hospitality or politics that starts with the calculation or controlling of these risks. However, if this calculation means the closure of all boundaries, not only territorial but also cultural, social and linguistic, this would mean the death of the nation. If the other by definition is incalculable, political calculations have to include a margin for the incalculable. In other words, Derrida (1997a: 13) refuses to close down hospitality to the logic of 'paternity' and (its extension the nation) or the logos because hospitality is the anti-logic of the logos, that is, of closure and determinism.

Hospitality, Postcolonial Migrants/Settlers and Colonial Legacy

Individual European countries have always been multiple and 'home to tensions between numerous religious, cultural, linguistic, and political affiliations' (Balibar, 2004: 5). The construction of the nation's 'imagined' political community has depended, among other things, on the 'historical insertion of populations and peoples in the system of nation-states' and hence on the nationalization of societies' cultures, languages and genealogies'. According to Etienne Balibar, this process has always been marked by the tension as well as reciprocal interaction between the double faceted construction of people as ethnos or the '"people" as an imagined community of membership and filiation' and demos, 'the "people" as the collective subject of representation, decision making and right' (2004: 8). This construction resulted in various subjective forms of the internalization of the border, or of the way people set imaginary borders between the groups they belong to (and hence develop a certain cultural nationalism). This construction also links certain democratic rights (right to education, social protection, security, and so on) with specific national belonging. What is interesting here is that the 'democratic composition of the nation led inevitably to systems of exclusion: the divide between the "majorities" and "minorities" and, more profoundly still, between populations considered native and those considered foreign, heterogeneous, who are racially or culturally stigmatised' (Balibar, 2004: 8).

These divisions were further aggravated by the history and legacy of colonialism and by the settlement in Europe of post-war immigrants from the ex-colonies. Balibar (1997: 391) raises the issue of how the idea of the 'Empire', with its hierarchical 'racial' differences that subordinate those seen as 'inferior,' still exercises influence in French society, for example. Thus, it plays the role of an interior frontier between the French and those who are carrying the image of the colonial subjects: the North African immigrants and their descendants. Balibar ironically points out that the image of young women – descendants of the ex-colonial North African Muslims in France – entering French schools with their Islamic headscarf[9] reminds the French that they have not just given up the Empire, but also of the failure of the supposed mission civilisatrice with its supposed aims to civilize and liberate colonial subjects from their 'beliefs' deemed as ignorant and primitive. In other words, Balibar questions the idea of the French Empire which was energized by 'prestige' and France's 'vocation superieure' (Said, 1994: 204) that believed in its ability to 'civilise' the colonial natives as factors that justified territorial acquisition. This is different from the British 'departmental view' as it is based on the French great assimilationist enterprise (Said, 1994: 204). However, French colonial assimilation that was supposed to start under the Revolution collapsed after theories of 'race' and 'racial inferiority' dominated French imperial strategies. Therefore, 'natives and their lands were not to be treated as entities that could be made French, but as possessions, the immutable characteristics of which required separation and subservience, even though this did not rule out the mission civilisatrice' (Said, 1994: 206). The colonial past, with its inegalitarian perception of native culture, casts its influence with the arrival of citizens from the ex-colonies who were assigned a 'subaltern' economic and cultural position as immigrant workers (temporary to permanent residency not acknowledged or recognized). This colonial legacy emphasizes the perception of 'legitimate' and 'deviant' cultures to cover for any social and economic domination.[10] Noiriel (1992) identifies this process of exclusion of certain nationals from equal rights as a form of 'tyranny of the national', which is due to the fact that these 'ethnic' groups are essentialized as inherently 'different'. One can clearly see this essentialization in the debate on the headscarf issue in France that started in 1989 and continued till the ban in 2004. Women came to be seen as the representatives of Islam,

9 Balibar (1997: 391–392) refers to the republican model of the school and the concept of secularity, which became politically charged during the 1989 headscarf affair when three Muslim girls were expelled from school for wearing the headscarves. This event marked the beginning of an intellectual debate about the principles of the Republic. Wieviorka (1992: 36) argues that it is not the principle of secularity that has caused the crisis, as teachers were used to seeing immigrant children wearing scarves and it was not perceived as a problem before then. But it became problematic at the end of the eighties because the immigrants themselves became a 'problem' threatening the cultural uniformity of the nation. In 2004, a law was passed to ban the display of 'conspicuous' religious symbols in public schools. This law mainly targets Muslims, particularly the wearing of the headscarf.

10 See Necira Guénif Souilamas (2000: 50) for an interesting analysis of how North African cultures in France, for example, are represented as 'deviant' and inferior' and how that perception covers up for the social and political exclusion of the immigrants and their descendants.

they are representatives of unacceptable differences that for some compromises the 'integrity' and 'cohesion' of the nation (Scott, 2005: 28). The debate as to whether this Muslim diaspora can be integrated into French society is still ongoing, while most, if not all of them believe that they already constitute a part of that society. The problem is that of refusing to recognize that the ideal of integration as belonging 'to a de facto historical and social entity' does not correspond to 'a mythical national type' (Balibar and Wallerstein, 1991: 223). Instead of tackling the real issues of social, political and economic inequalities that descendants of North African and African immigrants are confronting daily, the French Government has chosen to focus on the headscarf ban to depoliticize these issues and hence place the blame on its Muslim 'ethnic' minorities and not on the state's discriminatory policies. One must ask the question of whether the ban on the headscarf has brought any positive changes to French society, has it reduced social exclusion and racism? In fact, it has caused more resentment on the part of the French Muslims who feel under attack by the agents of the Republican institutions. As Vincent Geisser documents in his book *La Nouvelle Islamophobie* (2003), French Muslims (mostly of North African origin) have been the target of a growing tide of public hostility that has represented them as a threat to national security.

Most European countries seem unable to meet the challenge 'represented by the social, cultural, and political transition with which the presence of postcolonial and other sanctuary-seeking people has been unwittingly bound up. Instead, racist violence provides an easy means to "purify" and rehomogenize the nation' (Gilroy, 2004: 111). Anxiety about national identities produces defensive outbursts of nationalistic zeal. The rise of extreme nationalist parties in Europe raises concern about the resurgence of fascism whose past memory still haunts the continent. 'The immigrant problem' is a label under which all problems in Europe are justified: economic decline, loss of supposed national identity, insecurity and crime. The term 'immigrant' is also used to describe anyone who is not white, a fact that holds hostage to the process of immigration generations of non-white European citizens-settlers with strong bonds to the history and culture of their 'unhomely' homelands. The race-coded nationalism marked by the call for the purification of the nation from any outsiders/immigrants seems to provide an attractive and compelling option for confused and anxious Europeans. But, as Gilroy rightly asks, why have these anxieties generated 'racism and nationalism as the primary responses to problems that strangers and aliens import?' Why have the demands of postcolonial citizen settlers been met with latent exclusionary forces such as that of 'race' or ethnicity? (2004: 111). Gilroy explains this exclusion in terms of the inability of Europe to mourn its loss of empire and in terms of the 'fear at the prospect of open interaction with an otherness, which could only be imagined as loss and jeopardy' (2004: 111). The figure of the 'Asian' and Black immigrant in Britain and the North African in France, for example, is a reminder of the imperial past with all its failures as well as its successes. Descendants of post-war immigrants believe themselves to 'belong' to Europe, but they are still widely perceived as outsiders whose cultural, religious and social values will never be reconciled with 'European values'. This may be perceived as an attempt to turn postcolonial settlers into migrants long after immigration came to a halt.

Most European countries with a history of colonialism have failed to formulate a positive history of the post-war immigration of their ex-colonial subjects to the metropolis, an immigration that has contributed considerably to the prosperity of their countries. Therefore, the descendants of these migrant-settlers reject the memory of their parents' emigration as being linked solely to the mere economic logic that situates them in a subordinate position in relation to the role they played in helping European countries to reproduce themselves, this time not in the colonies but within the metropolis. In other words, they disavow the representation of their parents' history in Europe as 'guest labourers' chained to servitude and domination. They stress that their parents' displacement is historically linked to European colonial history and the decolonization process in their own countries. Hence, there is a strong need to re-think the history of countries like France and Britain, for example, in which colonial and post-colonial immigrant parents are seen as historic figures strongly linked to the imperial past and not as outsiders with no relation to their 'host countries'.

It is important to consider the way hospitality is shaped by the persistent legacy of colonialism. The continued exclusion of the post-war immigrants from the making of European history which still perceives them as an 'alien wedge' is a manifestation of the perpetuation of colonial culture or the 'empire within' that still preserves the same power structures that existed in the colonies (Young, 1990: 175). In a provocative way, Balibar claims that the emergence of a formal 'European citizenship' (or a system of institutions and rights common to various people linked to the European construction), is linked to the emergence of a 'European apartheid', defined as 'the critical nature of the contradiction between the opposite movements of inclusion and exclusion, reduplication of external borders in the form of "internal borders", stigmatisation and repression of populations whose presence within European societies is nonetheless increasingly massive and legitimate' (2004, x). This 'European apartheid' is not only about a category of foreign immigrant workers and asylum seekers being granted fewer rights, but it is more about the structures of discrimination in European nation-states 'that command uneven access to citizenship or nationality, particularly those inherited from the colonial past' (Balibar, 2004: 121). Given that there is a large immigrant population permanently resident in Europe, the appearance of structures of discrimination that are mostly inherited from the colonial past threatens to transform the former into aliens whose cultural difference is racially stigmatized.

It seems that Muslim diasporas, for example, represent a formidable challenge to the processes of inclusive nation building, pluralistic citizenship and a change of mode of thinking on Islam and Muslims in Europe. The increasing visibility of Islam and the demands of the Muslim diaspora for more rights are represented as a threat to the uniformity of the nation. Why cannot these demands be seen as signs of integration and not disintegration within European societies since they reflect the populations' strong feeling of being part of Europe, their homeland? Islam has been 'racialized' in Europe in the last decade and this 'racialization' fosters the implicit and latent exclusion of European Muslims from the benefits of constructive nation building and the full rights of citizenship.

European Muslims seem to be the heirs of Europe's long history of anti-Semitism. Islamophobia has the same characteristics as anti-Semitism since it draws on classical stereotypes and prejudice against Muslims and on physical and cultural differences that are blamed for disrupting the cohesion and stability of the nation (see Ben Jelloun, 1984: 85) Jews – considered for a long time as the other of Europe because of their supposed 'cultural', 'religious' and 'physical' difference – found themselves in the forefront of assimilatory ideologies. Their experience in European societies sheds light on the complexity of the Muslim diasporas' situation, as Bauman (1991: 145) argues, the Jews were seen as 'an admittedly unwieldy, scattered group spilling over any national border, they served everywhere as a symbol and a reminder of the assimilation's inner weaknesses, and, worse still, of the elusiveness of the dreamed-of-order'. The figure of the Muslim today serves as the social and cultural other (or 'stranger' in Bauman's words) of Europe. If the figure of the Jew stood against the idea of a supposed collective identity, the figure of the Muslim disrupts the cultural conformity of the nation. In France, for example, 'new conditions of acceptance were to be unashamedly self-canceling: a Jew could become a Frenchman only if he was a Frenchman; that is, if he was not a Jew. The states of being a Jew and being a Frenchman were declared mutually exclusive – neither stages of a life-process, nor two faces of the same identity' (Bauman, 1991: 152–153). French citizens of North African Muslim origin are subject today to the same rhetoric of 'foreignness' that portrays them as the other of France or the 'almost-close-to-French' but that can never be French. The more they reveal their proximity, the more they are kept at a distance. Therefore, this vicious circle of assimilation and difference reveals that the nation is not the product of learning and self-improvement but in fact it is a 'commonality of fate and blood – or not a nation at all' (Bauman, 1991: 154). Bauman argues that the function of the stranger is to draw clear cultural and social boundaries between those seen as natives and those seen as aliens (1992: 683). It is the figure of the Muslim today that serves to reinforce such internal boundaries since s/he is seen as the cultural other. However, as Bauman (1991: 55) argues, the stranger also calls fixed boundaries into question and hence destabilizes them since s/he falls in-between the national. In other words, the stranger problematisises such fixity because s/he can move across cultures and hence emphasizes the uncertainty and fluidity of those social and cultural boundaries. Moreover, if the diasporic 'hybrid' Jew with his insider/outsider status has contributed to the flourishing of modern culture (Bauman, 1991: 154), postcolonial diasporic Muslim and non-Muslim populations are enriching the social and cultural life of Europe with their diverse cultural and artistic productions.

A revision of the history of past models for the formulation of rights may be a good starting point towards a reformulation of the concept of citizenship, and also the perception of difference in a non-essential way but as a process that implies constant transformations. The aim is for a citizenship that will not be exclusionist and will be free from the straitjacket of national belonging and thus open to flexibility and change, and responsive to the negotiation of common and particular interests (Silverman, 1995: 261). In this historical moment, it is necessary for Europe to collectively invent a new image of its population, 'a new image of the relation between membership in historical communities (ethnos) and the continued creation of citizenship (demos) through collective action and the acquisition of fundamental rights to existence, work and expression, as well as civic equality …' (Balibar, 2004: 9).

Conclusion

Derrida introduces the absolute irreducibility between the ethics of unconditional hospitality, which is based on the absolute welcome of the Other without any restrictions, and the politics of conditional hospitality, which is based on the restrictions of law making. Even though the *hiatus* between the ethics (the law) of hospitality and the politics (the laws) of hospitality exists, the two cannot exist separately. This *aporia* does not mean paralysis, but in fact, it means the primacy of the ethics of hospitality over politics, and thus, keeping alive the danger of hostility in the making of the politics of hospitality by 'political invention' that respects the uniqueness of the other every time a decision is taken.

Derrida stresses that neither hospitality nor ethics can exist without politics or democracy and vice versa. Democracy, like hospitality is marked by the same aporia between the law and the laws, between incalculation, unconditionality and calculability, conditionality. Derrida suggests the idea of democracy-to-come that would free the interpretation of the concept of equality from its 'phallogocentric schema of fraternity', which has dominated Western democracies. The concept of fraternization has played an important role in the history of the formation of political discourse in Europe, especially in France. Such a democracy would be 'a matter of thinking alterity without hierarchical difference' (Derrida, 1997a: 232). Democracy-to-come has the character of 'the incalculable', like that of unconditional hospitality, but its incalculability, that resists 'fraternization', or the tribal and the national, allows the calculability of politicization and thus ameliorates the existing democracy. It is an opening of democracy beyond the juridical and towards a space where the juridical and the ethical can intersect, where the law and laws of hospitality could uncomfortably and paradoxically cohabit. It is a form of 'providing constant pressure on the state, a pressure of emancipatory intent aiming at its infinite amelioration, the perfectibility of politics, the endless betterment of actually existing democracy' (Critchley, 1999: 281).

If post-war immigrants in Europe were considered for a long time as a temporary foreign labour force and thus had to be kept outside political and social affairs, the most recent realization of their settlement in the host countries has given rise to a 'sociological approach' that still grants them a marginal place in society. The immigrants and their descendants are used to 'strengthen' the coherence of the main community and thus reinforce the dialectic of proximity and distance, which situates the immigrants and their descendants (who are European citizens) in a position of social foreignness and territorial exteriority. Moreover, the cultural specificity of Europe's postcolonial diaspora has been constructed in terms of the 'double culture', that is, a culture that cannot integrate with the European one (especially Muslim cultures) because of their irreducible differences. The emergence of Islam in the public sphere has made Islamic rituals visible and thus has raised the idea of its incompatibility with Europe's 'secular' values. Therefore, hospitality is not only marked by the 'autochthonous', the 'familial' and the national that exclude the other, but it is also marked by the legacy of colonialism with its hierarchical and racist subordination of other cultures and people. Descendants of postcolonial migrants still carry the image of the ex-colonial 'immigrant' with its violent colonial residue that relegates them to the margins of society on the basis of their 'cultural', 'ethnic', 'religious' and social affiliations that are sometimes deemed incompatible with European values. The history of post-war migration to Europe must not be limited to the crude economic perspective (Europe's need of a labour force after the Second World War) because that denies the historical complexity of colonialism and postcolonialism. The history of immigration is part of the imperial history of Europe. With their mixed origins and cultures, descendants of post-war immigrants can resist monolithic representations of cultures and histories and can suggest new alliances and solidarity that transcend skin colour and thus open hospitality beyond nationalistic and ethnic determinism.

References

Balibar, E. (1992), *Les Frontièrs de la Démocratie* (Paris: La Découverte).

—— (1997), *La Crainte des Masses* (Paris: Galilee).

—— (2004), *We, the People of Europe?* (Oxford and Princeton: Princeton University Press Books).

Balibar, E. and Wallerstein, I. (1991), *Race, Nation, Class: Ambiguous Identities* (Paris: La Découverte).

Bauman, Z. (1991), *Modernity and Ambivalence* (Cambridge: Polity Press).

—— (1992), 'Soil, Blood, Identity', *Sociological Review*, **40**(4), 675–701.

Ben Jelloun, T. (1984), *Hosptialité Française: Racisme et Immigration Maghrébine* (Paris: Seuil).

Critchley, S. (1999), *Ethics, Politics, Subjectivity: Essays on Derrida, Levinas and Contemporary French Thought* (London and New York: Verso Books).

Derrida, J. (1997a), *Politics of Friendship*. trans. Collins, G. (London and New York: Verso Books).

—— (1997b), 'Quand j'ai entendu l'expression "délit de l'hospitalité"', *Plein Droit*, **34** (April), 3–8.

_____. (1999a), *Adieu To Emmanuel Levinas*, trans. P-A. Brault and M. Naas (Stanford, CA: Stanford University Press).

—— (1999b), 'Débat: une hospitalité sans condition' in Seffahi (eds).

—— (1999c), 'Une hospitalité à l'infini' in Seffahi (eds).

—— (2000a), 'Hospitality', *Angelaki*, **5**(3), 3–18.

—— (2000b), *Of Hospitality: Anne Dufourmantelle Invites Jacques Derrida to Respond*. trans. Bowlby, R. (Stanford, CA: Stanford University Press Books).

—— (2001), *On Cosmopolitanism and Forgiveness* (London and New York: Routledge Books).

Geisser, V. (2003), *La Nouvelle Islamophobie* (Paris: La Découverte).

Gilroy, P. (2000), *Between Camps: Nations, Cultures and the Allure of Race* (London: Penguin Books).

—— (2004), *After Empire: Melancholia or Convivial Culture?* (London: Routledge Books).

Guénif Souilamas, N. (2000), *Des 'Beurette' aux Descendants d'immigrants Nord-Africain* (Paris: Éditions Grasset and Fasquelle).

Hargreaves, A. and Leaman, J., eds (1995), *Racism, Ethnicity and Politics in Contemporary Europe* (Camberley, Surrey: Edward Elgar Publishing Limited).

Kant, I. (1957), *Perpetual Peace* (New York: The Library of Liberal Arts).

—— (1974), *Anthropology from a Pragmatic Point of View*. trans. Gregor, M.J. (The Hague, The Netherlands: Martinus-Nijhoff Books).

Levinas, E. (1961), *Totalité et infini: Essai sur l'extériorité* (The Hague: Martinus-Nijhoff Books). [Trans. A. Lingis as *Totality and Infinity: An Essay on Exteriority* (Pittsburgh: Duquesne University Press).].

Noiriel, G. (1992), *Population, Immigration et identité nationale en France – XX. Siècle* (Paris: Hachette).

Rydgren, J. (2005), 'Is Extreme Right-Wing Populism Contagious?, Explaining the Emergence of a New Party Family', *European Journal of Political Research*, **44**(3), 413–437.

Said, E. (1994), *Representations of the Intellectual: The 1993 Reith Lectures* (London: Vintage Books).

Scott, J.W. (2005), *Parité! Sexual Equality and the Crisis of French Universalism* (Chicago and London: The University of Chicago Press).

Seffahi, M., ed. (1999), *Manifeste pour l'hospitalité, aux Minguettes: Autour de Jacques Derrida* (Gringy: Paroles d'aube).

Silverman, M. (1995), 'Rights and Difference: Questions of Citizenship in France' in Hargreaves and Leaman (eds).

The Economist (2000), 'Go for it: Europe needs more immigrants', 6–12 May, 25–31.

Wieviorka, M. (1992), *La France Raciste* (Paris: Seuil).

Young, R. (1990), *White Mythologies: Writing History and the West* (London and New York: Routledge Books).

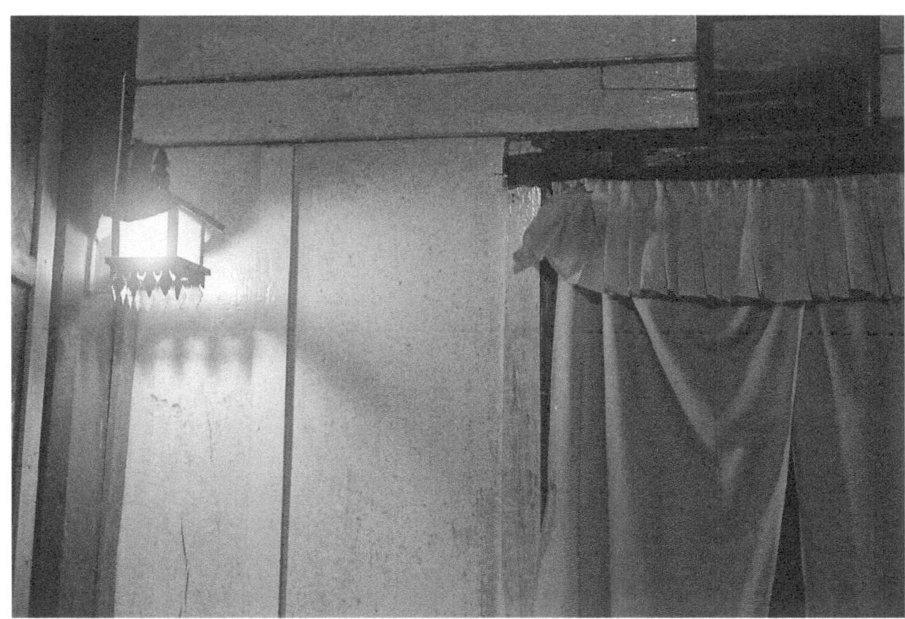

Pink Curtains (Luang Prabang, Laos, 2004) by Elly Clarke

Chapter 11

Figures of Oriental Hospitality: Nomads and Sybarites

Judith Still

In France today, even more than in other European nations, public discussion about post-colonial immigration and refugees often involves reference to the code and practice of *hospitality*.[1] Discourse on hospitality frequently refers explicitly or implicitly to a past time period when conduct is deemed closer to the essence of generosity or reciprocity. For example, the Republican tradition dating back to the French Revolution, with its exaltation of the 'droits de l'homme' (apparently gender-neutral in the English 'human rights') is evoked today, in a mythical fashion, as a model of universal fraternal hospitality (see Fassin, Morice and Quiminal, 1997; Gotman, 2001; Rosello, 2001). Sometimes cultures which are spatially and/or culturally distant, for instance the Arab culture of nomadic hospitality,[2] are used as a proxy for temporal distance. The intertexts of these other locations function rhetorically in the analysis of hospitality here and now, and, more specifically, of the *failure* of hospitality today (whenever and wherever today is). Myths of Arab hospitality, or inhospitality, are particularly potent in a country where unofficial and official racism is directed more at those who are perceived as 'Arabs' than at any other group (MacMaster, 1997). In this chapter, alongside evoking the complexity of present-day reflections on hospitality, I shall take the example of early Enlightenment writing about Oriental hospitality and the contrast with French inhospitality in the same period.

---

1 I am engaged in a study of hospitality in eighteenth-century France and the contemporary period funded by the Leverhulme Trust, to whom I should like to express my gratitude (see Still, 2003, 2004, 2005a, 2005b). The present article was given as a paper in various different forms, first at the IALIC conference in Brussels in 2005, and then at the London French seminar and the *Café Scientifique et Culturel* in Nottingham in 2006. I am grateful to the various audiences for the contributions they made to my thinking about these questions, and to Ahmad Gunny for reading the paper.

2 I shall give two examples, both in the context of a critique of contemporary French inhospitality towards immigrants or refugees. Ben Jelloun (1999) writes in praise of Moroccan hospitality as part of Moroccans' tradition and cultural identity. In Hélène Cixous's programme notes to *Le dernier Caravanserail* (see note 9), a Sufi tale of Hospitality is reproduced. It begins by telling how the people of Turkestan are famed for their generosity, pride and love of horses; and ends with the impoverished Anwar Beg feeding his beloved and extremely valuable horse to the guest who had wanted to buy it from him, saying: 'hospitality comes before everything else'.

Contemporary Interest

The renaissance of hospitality as a topic is related to recent movements of population towards Europe, immigration and notably the arrival of asylum seekers, who have been figured as guests; and to the political responses in Europe to these new comers, which have been represented and analysed not only by political scientists or sociologists, but also from the arts, as hospitality or inhospitality (Gotman, 2001; Rosello, 2001). A second factor in this renaissance is the existence of a growing body of powerful philosophical writing on hospitality (Jacques Derrida's work is a key example) some of which pre-dates the current wave of post-colonial xenophobia, and most of which draws on experiences of colonialism and of the Second World War as well as more recent events. The third factor, which is perhaps more powerful in the UK than in France, but is important throughout the world is commercial globalization, tourism and travel. At first glance, hospitality may seem a matter of inviting friends or relatives to your house, but, beyond moral and social relations between individuals, hospitality can be, and *is*, evoked with respect to relations between different nations or between nations and individuals of a different nationality. The personal dimension that we experience in our everyday lives both inflects our *analysis* of the political dimension and is *part and parcel* of it for many migrants who find a (temporary) home with kin, friends or acquaintances.

Reference to the Past

Although there has been a recent resurgence, hospitality is a topic that has consistently been considered important over long periods of time, and over wide tracts of the globe. Our conviction of its *universality* is indeed critical to our understanding of its structure: hospitality is traditionally defined as a universal (even *the* universal) human virtue. As Derrida summarizes it in *Adieu to Emmanuel Levinas*: 'For hospitality is not simply some region of ethics, let alone [...] the name of a problem in law or politics: it is ethicity itself, the whole and the principle of ethics' (1997, 50). Derrida's work on Levinas and hospitality suggests the complicated force of that copula 'is'.

Of course, according to most rhetorical evocations, hospitality is practised more faithfully at certain times or by certain peoples. I say human virtue – hospitality is traditionally offered and accepted by men, and in some contexts is what reveals the virtue, the manliness, the humanity of the man (although women very often perform the labour of hospitality). Nevertheless, like most other forms of human relationship, its significance varies to some extent over time and space. Different cultures have different modes of hospitality, and, as we look to the future, we should think about constructing new modes suitable to a new historical moment. One of the things that interests me about our formal and informal discussions of hospitality is their *intertextual quality* – how elements from a range of earlier or otherwise distant theories and practices are introduced and transformed in the present. How we are haunted by the past, and how we fashion those ghosts in the present. The debate in France draws on a multitude of textual strands, including Homeric hospitality, biblical hospitality, pastoral traditions, and of course the mythical political hospitality of the French Revolution. The reference to the French Republican tradition often echoes rhetoric of the 1790s, whereby the State would welcome those lovers of liberty and equality who suffer under totalitarian regimes (such as late eighteenth-century England). The work of historians such as Wahnich (1997) shows how the discourse and practices of the 1790s were already self-contradictory. She analyses certain key discourses such as Condorcet's address of December 1791 to be sent to foreign peoples assuring them that 'the principle of hospitality' will not be put in question by the war – deserters from foreign armies can become French citizens. In fact, foreigners quickly become assimilated into the category of spies or counter-revolutionaries, and Wahnich points to the decree of September 1793 that expels foreigners born in those countries at war with France. She also follows the fates of certain famous foreigners in France such as Anacharsis Cloots, who points out what a barbarous expression the very word '*étranger*' (foreigner or stranger) is. One of the reasons for speculation on the past is to discover fractures or contradictions in what may be *presented* as a seamless or utopian narrative.

I should add that the rhetorical triangle of self-critique ('we ought to be doing better just as they did') I have identified here, between ourselves or our nation as *hosts*, others or strangers who are potential *guests*, and those who were or are *better hosts*, is not the only critical triangle. Another relevant structure is the self-justifying comparison ('we're not as bad as them') so often drawn between ourselves in our real or potential relation to subalterns (for example slaves or colonized peoples) and other nations, sometimes competitor nations with respect to their relation to subalterns. For instance Spain and Portugal are often cited as examples of bad colonizers because of their notorious practices in the Americas. I was amused by a work defending Montesquieu from the charge of Orientalism on the grounds that he condemned Spain and Portugal for their cruelty in the New World, sees faults in the French Old Regime and so on.[3] While this kind of condemnation is often sincere it can serve as exculpation for the speaker as *colonisateur de bonne volonté* (well-intentioned colonizer) as can the rhetorical figure Barthes (1973) calls inoculation: introducing criticism of what seem minor contingent problems to ward off radical overthrow of the system. In this article I shall focus on two 'others' to present-day France: the late-seventeenth to early-eighteenth century and the Orient.

Nomads and the Caravenserai

In order to give a flavour of French eighteenth-century representations of the Orient I could go to a range of well-known authors, but in this chapter I wish to propose a

3 Schaub cites Montesquieu's writings on the New World as proof that he is 'a most determined foe of imperialism' (1995: 7), and therefore cannot be accused of Orientalism. She believes that 'new academic fashions' (8) are either 'like deconstruction [...] in principle hostile to the text' or 'like feminism' use a 'litmus test' (8), and so, 'Judged by the anti-imperialist litmus test, Montesquieu (despite being a dead, white, European male) turns the right color' (7). But of course it is perfectly possible to be both a foe of the evils of [Spanish] imperialism and an Orientalist.

neglected figure as worthy of our attention: the diamond merchant Jean Chardin, who, as Agha Chardin represents the traveller, and as John Chardin, the Protestant refugee.[4] While Chardin (1643–1713) may be cited as a seventeenth-century phenomenon, his major publication, *Les Voyages en Perse* which recounts his travels to Persia in the 1660s and 1670s, only came out, just a couple of years before his death, in 1711[5] and had a significant influence on many Enlightenment thinkers – not only the obvious Montesquieu in *Persian Letters* (1960), who will provide a counterpoint to Chardin throughout this article.[6] He contributes largely to what will become Orientalism; as Edward Said points out, 'every writer on the Orient (and this is true even of Homer) assumes some Oriental precedent, some previous knowledge of the Orient, to which he refers and on which he relies' (1978: 20). Study of Chardin's authoritative work reminds us that that disciplinary designation encompasses knowledge, translation, affection, admiration – it is not only or not always a peddling of negative stereotypes although it does represent the Orient and speak on its behalf.[7] Orientalism draws from,

4 Chardin, travelling as a merchant, adopts the language, customs, dress and food of his hosts – the translation of his name as 'Agha' is an emblem of this practice, indicating not only how he was addressed by his hosts but also how he was perceived back in Europe where he was often designated 'the Traveler' and represented in Persian robes. When he takes refuge in England he is also known as 'John', and indeed his surname is often anglicized as well, or Sir John once he is knighted – searching in catalogues today we need to take account of this.

5 The first volume was first published in 1686 in London and Amsterdam; it was revised and expanded in the 1711 publication. The best edition to consult is perhaps the 1811 one edited by L. Langlès as this restores some of the cuts made by the publisher in 1711 – for fear that those observations seemed anti-Catholic. Very little has been published on Chardin; the most comprehensive account of his works as well as his life is Van der Cruysse (1998) which I have found very useful. I should thank Chris Johnson and Michael Moriarty for lending me their copies of two modern editions of Chardin, both sadly now out of print. My translations throughout unless otherwise indicated.

6 Vernière, (Montesquieu 1960) locates 38 certain references to Chardin (and nine possible or probable references) in the *Lettres Persanes* (xx), No other source is anything like so frequent.

7 Douthwaite (1992) claims: 'No matter how sympathetic his portrayal of the Persians, however, Chardin's effort to catalogue and distill the strange practices of this "other world" for the European reader ultimately has the same effect as the "informational" discourse of later travel writings; it reinforces the reader's sense of being privy to an extraordinary spectacle, frozen in time and safely distanced from the dynamism of European life' (79). She goes on to cite Talad Asad on 'the early Orientalists' and their descriptions of an 'unprogressive' and 'fanatical' Islamic state which offered an ideal pretext for European efforts at commercial expansion and political domination in the Levant. In fact early Orientalists such as Chardin (coming after the anti-Islamic period of the Crusades but before the high colonial period when racism is institutionalized), and Montesquieu in his wake, present Islam as tolerant – unlike Catholicism. And Chardin had experienced at first hand the fanaticism of Catholicism when he wrote *Voyages en Perse*. See Gunny (1996) for the argument that, while the late seventeenth and eighteenth centuries were not immune from what preceded them nor from the beginnings of imperial designs, texts such as those of Chardin were indeed relatively 'enlightened' in their depiction of the Orient in general and Islam in particular.

and contributes to, the discursive construct of 'The Orient' – of course Orientalist texts are often hybrid and contain elements which are positive or scientific.

Representations of Oriental hospitality are typically bipolar: they focus on either simple nomadic and rural practices or luxurious entertainment. The former is more obviously suitable to be set up as a model of past generosity; the latter is more obviously ambiguous. However, I should note in passing that almost all these idealized practices from long ago or far away exclude or repress women, and that exclusion/repression echoes with us here and now. Kaplan (1996), amongst many others, attacks contemporary theory and other cultural artefacts that use notions of travel, or mobility, or the nomad, in an ahistorical and universalized fashion, the most famous example being Gilles Deleuze and Felix Guattari's 'deterritorialisation' (1980). I shall not repeat these arguments; Peter Hallward's recent book on Deleuze (2006) already does a fine job of showing the metaphysical underpinning of 'nomad thought'. Instead I wish to take one specific and located historical instance (which has been enormously influential in the construction of modern Orientalism) of the representation of travel in the Orient. I shall begin with the Caravanserai as a key example of Oriental hospitality available to all strangers with little regard for rank or wealth. Accommodation in caravanserais is characteristically basic even though the architecture of some of the great stone caravanserais is described as very beautiful: Chardin provides an illustration of the one at Kashan which, he says, is the loveliest in Persia. The caravanserai establishes a shifting community of travellers. It is an emblem of mobility even though European visitors suggest that such a system relies on a basically static society (see Montesquieu, 1960: i, v).

The relative mobility or immobility of different nations or different historical periods is a complex question. There is an Early Modern *topos* that the Parisians, for example, are excessively mobile in a number of senses. In *The Persian Letters*, Rica writes: 'they run; they fly. The slow vehicles of Asia, the measured step of our camels, would give them apoplexy' (Montesquieu, 1973: 72). Vernière notes that this is also a theme in *L'espion turc* or in texts by Cotolendi, Dufresny or Boileau (Montesquieu, 1960: 55). Today, we assume that our postmodern globalized society is more mobile than earlier periods. Yet this assumption often pulls together quite disparate issues: the mobility of the migrant or refugee who makes one long and traumatic journey in his or her life (in any century) is different from that of business men or of tourists – even though one person (like Chardin) can embody different kinds of mobility at different moments in his or her life. Refugees, soldiers, pilgrims and migrant labourers existed in considerable numbers before our own time. The democratization of tourism is clearly a post-modern phenomenon but we should remember that it is not universal – there are still parts of the globe and still pockets even within wealthy West European nations where being a tourist is not a possibility, and there are gender differences to be considered as well as questions of class, nationality or race. I shall now go on to quote some of Chardin's accounts (1811 edition) of caravanserais, focusing on three points he emphasizes: mobility, community and simplicity.

Mobility

'Caravanserais are large buildings constructed to shelter travellers. You must realise that in Asia you do not see nearly so many strangers in towns or on highways as there are in Europe' (II, 142).

Chardin suggests that you see fewer strangers in Asia, and so there are fewer commercial inns or hotels, for three reasons: because it is less populated than Europe; because the climate is more favourable and so there is less need for activity or trade; and because women must not be seen by men which makes travelling as husband and wives, or as a family, rather complicated.[8] While there is less actual mobility, according to Chardin, there is an assumption of, and provision for, mobility. The word *caravanserai* itself is made up of the word for travellers (or 'returners' as they are known for good luck, Chardin tells us) and the word for palace or *serail* (bringing the listener or the reader back to the harem in our imaginations if we so wish). This combination is supposed to remind all that they are but travellers on this earth even if they inhabit a great palace. Chardin relates various Persian stories to illustrate this maxim. The caravanserai in Kashan (Cachan in Chardin's spelling), an area made rich by the manufacture of silk and other fabrics, was built by Abbas the Great, who put in the inscription:

> The world is a caravanserai, and we are a caravan.
>
> In a caravanserai, do not build caravanserais.

Chardin, having translated this into French for his readers, also glosses it: 'In other words, we should not bank on a stable and solid dwelling in this world which is only somewhere we pass through.' (III, 4.)

Simplicity

Caravanserais do not provide equipment for travellers such as bedding or cooking utensils – these have to be transported by the traveller – they simply provide shelter. However, as Chardin points out, Persians do not feel the lack of tables, chairs or bedsteads; their habits are different and quite appropriate for the climate.

> The only thing that you find in this kind of accommodation is the four walls. Each person, as he goes in, takes the first room he finds empty, on the side *he likes*. He stays so long as *he likes*, and then leaves without anyone asking anything of him. Rich folk give the concierge's servant a few pennies as they leave, as much as *they like*; for you cannot be asked to pay rent as these buildings are good works, as they say, in other words charitable foundations to serve travellers, where the concierge and servants are paid to take care

8 See *Persian Letters*, I.I. for the complexity involved in women travelling. One of Usbek's wives, Zachi, writes: 'What troubles journeys cause for women!' (XLVII, 103); this seems bathetic understatement in the context of her report that two men were killed while the wives cross a river – all for the sake of 'virtue'. The editor cites the influence of Chardin e.g. III, ***57***, *II* (ch. 13), 281, as well as Tavernier.

to them. The concierge usually sells what is necessary for the horses and the most basic provisions such as bread, wine… (II, 145–6, my emphasis).

I emphasize the repetition of 'il lui/leur plaît'. Entertainment often involves pleasing or pleasure; here *plaire* is used in its simplest sense. It signifies free-ness: liberty from gratuity, *freedom from* the constraint of any cost (financial or social) and thus freedom to choose, as opposed to either the aristocratic *freedom to* abuse or the freedom of the market (exploitation). True hospitality means the freedom of the house. This simplicity is above all true of rural caravanserais; in big cities things are a little more complicated of course (urban hospitality is usually coded differently from rural hospitality). In cities, as well as free accommodation for travellers, there are special commercial caravanserais for merchants that are a little more luxurious (and 'having doors to the rooms that close firmly' [I, 146]). Urban caravanserais were also *vulnerable* to prostitutes, as Chardin would put it. Closed doors create other openings (both on an individual and a national level – border controls create a market in people trafficking). The different urban caravanserais are specific to trades or nationalities; if you want to buy Indian fabrics, for example, you know where to go – or if you want news of a certain place or a person of a certain nationality.

Community in Irivan, Chardin is asked by the Governor: 'where I should prefer to stay, in his fortress or in the caravanserai he had built [...] I chose the caravanserai, because there is no safer place, and because you never lack for company, as there are merchants from throughout Asia' (II, 194). The caravanserai creates a temporary community of (male) travellers who can give each other news and tell stories of their travels. As Hélène Cixous will say much later: 'In the caravanserai, travellers (as has always been the case ever since Odysseus) become story-tellers' (*Le dernier Caravansérail* programme notes).[9] Men create caravanserais because of belief in a fundamental human commonality – we are all transients on this earth. Everyone should have a right to shelter for a time. A vertical or transcendental faith in God (here, Islam) may underpin these practices but it is horizontal relations – fraternity – that is foregrounded.

The caravanserai, as described by Chardin, and as taken up by many later writers, thus brings together three of the elements (mobility, simplicity and community) which tend to be positively valorized in discourses on hospitality. There is no essential link between these three aspects, but it is not surprising that they are frequently brought together when writers, reflecting on the possible models of welcoming strangers, turn to rural or nomadic practices. When travellers are perceived as moving from place to place, rather than making one journey and settling (as Chardin the refugee will do), then simple, free (without ties) accommodation which permits transient and shifting communities to be established and dissolved is often celebrated. The

9 Cixous collaborated with Ariane Mnouchkine in the creation of a production staged at the Théâtre du Soleil in Paris entitled *Le dernier Caravansérail. Odyssées* (The Last Caravanserai. Odysseys), which was still running in February 2006. Actors speak the words, the stories, of refugees from a number of 'Eastern' countries (including what was formerly known as Persia) who, as she puts it 'stay for a short while in one caravanserai or another' (Programme notes, n. p.). In the programme notes Cixous refers repeatedly to Derrida's work on hospitality.

caravanserai is not of course the only model of past (or spatially distant) hospitality to be evoked – I have already mentioned Republican revolutionary hospitality as another example, and one which would theoretically provide a more permanent solution to 'the problem' of the foreigner. The Homeric model of the guest-friend (or *xenia*) is another example of a schema for mobility rather than for the permanent incomer.

Sybarites

I should like briefly to consider the sybaritic pole as well as the nomadic pole of representations of Oriental hospitality. Space constraints will mean that I do not treat you at length to any of the possible descriptions of banquets and other kinds of luxurious hospitality in Chardin's writings – his first publication was *Le Couronnement de Soleïmaan*, describing the magnificent coronations of Suleiman. I shall merely note that luxurious aristocratic hospitality relates to what I call the urban (clearly that term needs glossing according to period and culture), and that it has a tendency to conjure up commerce, cruelty and lasciviousness, haunted by the harem, sérail or seraglio (see Alloula, 1981 and Grosrichard, 1979). 'Commerce' may surprise when we consider the notion we have, in Georges Bataille amongst others, of aristocratic *dépense* opposed to bourgeois economy. Chardin is particularly struck (and sometimes exasperated) by the complexities of the relationship between commerce and gift-giving or hospitality in Persia – which makes his role as a trader less straightforward than he would wish. While we may note a cultural difference with regard to the rituals of exchange, we may also note our tendency to project onto the Orient the underbelly of our own self-image. The Sun-King, with his ostentatious expenditure, was not immune to miserly or deceitful practices when he owed money – and nobles could follow suit.

One of the more famous accounts of luxurious entertainment is that of the reception of ambassadors by Shah Suleiman. (Chardin uses the terms 'King' and 'Shah' interchangeably.) This description (like many others Chardin gives) highlights the importance of bearing suitable gifts when you are entertained, and thus the interpenetration of hospitality and (economic) exchange. Readers learn about foreign customs not only out of intellectual curiosity or pleasure, but also sometimes with a practical purpose such as trade (see Still (2006) on *Du bon usage du thé*). Persian rulers had complicated ceremonies for the reception of gifts from foreign visitors. There is a tension between the incalculable: presents are ultimately to be brought in from the East for religious reasons (all gifts are from God); and calculation. Gifts to the Shah would be distributed to the chief bourgeois in the city who would each have selected local shopkeepers to deliver them some days later to the King: 'They do this both for the honour of the person making the gift, because it makes him look more impressive, and for the grandeur of the King, since the peoples, seeing the gifts brought to him, reckon that he is highly valued by foreign nations.' (II, 170.) Chardin assures his readers that the King is so respected that even if 600 people are involved in this spectacular display nothing is ever lost. Before the presents are stored, their value is estimated and recorded as, in addition to the presents themselves, the donors also have to give to the King's officers the equivalent of 25 per cent of their value. This is a delicate calculation as an over-estimate involves financial loss, but may benefit the donor's standing at court. Chardin tells us that most presents originating in Europe (such as mirrors, pistols or pictures) are badly chosen without sufficient awareness of cultural differences, and so simply gather dust over the centuries in the Shahs' warehouses.

He describes both a reception for Muscovite ambassadors and then the repetition five days later for French and English guests. These are remarkable both for the extraordinary luxury which he details with his jeweller's eye: beautiful horses with harnesses encrusted with diamonds, rubies and emeralds; lions, tigers and leopards eating from golden bowls; wonderful food served in gold vessels studded with precious stones; gladiatorial displays and so on; and also for the inhospitality on a personal level. The Ambassador is brought to prostrate himself before the King and hands his letter to a captain who passes it to the Prime Minister who gives it to the King – who puts it down without even glancing at it. In the course of the reception, the King never speaks to, or even looks at, the Ambassadors nor does he deign to look at their gifts. Chardin notes, however, that his father (Abbas II) was more affable, and graciously engaged guests, including himself, in conversation. This is significant as Chardin's accounts of Suleiman's behaviour (which represent a distinct decline in standards of royal courtesy for Chardin) are sometimes cited as the epitome of Western Orientalism. Grosrichard (1979) makes an interesting case, for instance, that the model (or spectre) of absolutism derived from Chardin's Orient is structured by the invariable spectacle of the blinding despot, who must be obeyed instantly and in silence. Yet even as we remark how Chardin's descriptions will feed into, for example, the cliché of 'le faste oriental' – given even by a standard modern French dictionary (Le Petit Robert) as a gloss on 'faste' or luxury – it is also important to note that his long study is not seamless. The Persian court is not static; it changes over the time of Chardin's two lengthy visits (roughly two and five years respectively), and he notes all sorts of distinctions between practices.

As well as precious objects, food and drink, Chardin is interested in the particular forms of commoditization of women in Persia. Dancing girls, who feature significantly in entertainment, receive wages as well as 'presents', and are often named by their price (II, 210). In Ispahan there are also huge numbers of both registered and unregistered prostitutes who are, Chardin informs us, very expensive. This surprises him:

> When you consider that in Persia, on the one hand, the religion allows anyone to buy slave-girls, and to have as many concubines as they like: which should bring down the price of prostitutes; and, on the other hand, young people do not have a lot of money and are married off early. We must attribute the cause to sexual desire which stings more sharply in these hot countries than elsewhere, and to the art of these creatures which is a kind of magic (II, 212).

Supply may be abundant, but demand is even greater. Many a young nobleman is ruined by enslavement to a bewitching prostitute who sends him packing after she has spent his last penny, Chardin informs his readers. 'You recognise these slaves to love by the burn marks on their bodies', the more self-inflicted brands the more the man is seen to be in love.

We Westerners are fascinated by pornographic clichés of the harem – 'not us' – yet our very fascination indicates a place for that 'not at all us' within ourselves.[10] Montesquieu catches his reader from his second Persian Letter onwards before he turns to social and philosophical speculation – and *Letters* III and IV are particularly titillating in their descriptions of Usbek's harem.[11] By the end of the novel, for example Letter LXXIX the reader can combine voyeurism and sadism – imagining the stripping naked and beating of supposedly intensely modest women.

To turn to France today: the problem with sexual repression and violence in the North-African dominated *banlieues* (housing schemes on the outskirts of large French cities) has received a certain amount of media publicity; we can sit comfortably and read with a certain *frisson* about gang rape euphemized with the term '*tournante*' or women cruelly punished for transgression (Amara and Zappi, 2004: 54). Can we continue to expel these shameful practices – an internal exile in the ghettos – or should we face up to our part in the fate of many migrants '*sans papiers*' (undocumented and hence potentially illegal), including the trafficking of women from Eastern Europe. What about our part in the guilty secret of even middle-class women, *bourgeoises*, at home sometimes confined, sometimes battered, sometimes raped by their husbands albeit *Français de souche* (white Frenchmen)? Brothels or red-light areas in European cities today (never mind pornography whether on the net or in other forms) could be seen as the democratization of the harem – democratization meaning, as it so often does, an expansion to include larger numbers of men. Up till recently most British prostitutes were British nationals, now it is said that about 80 per cent of sex workers in the UK are foreigners. And swelling supply means prices have fallen. The pride of the Persian harem was the Circassian

10 Kabbani (1986) makes a couple of references to Chardin selecting quotations to demonstrate a fascination with the 'cruel and vengeful Eastern male' and his sexual victims. She claims that his 'writings on Persia were instrumental in the forging of the eighteenth-century's views of that part of the world. Although a sensitive and studious traveller, Chardin never managed to be an impartial one. He could not succeed in eluding the obligatory *topos* of the seraglio which Europe held so dear. Chardin emphasized the severity prevalent in the seraglio, enumerated the restrictions against women, provided examples of the capricious punishments that they were obliged to endure' (26). According to Kabbani, Chardin was also obsessed with tribadism in the harem and other sexual 'perversions' in the Orient. She links him to the epitome of exoticism: Antoine Galland, the translator (and inventor) of the *Mille et une nuits* (26–27) and Sir Richard Burton, the translator of *Arabian Nights* (58). While there is the brief parenthetical 'sensitive and studious', this thumbnail sketch of Chardin's work and influence is extraordinarily one-sided – although perhaps necessarily given the focus of her book.

11 While it is clear that the titillating or 'Gallic' quality of the harem letters is an element both in seducing the 'Gallic' reader and in 'disgusting' those who sought to ban the book in the nineteenth century for example, the letters also lend a voice to the cultural and sexual other. Douthwaite (1992) argues that Usbek and Rica become assimilated as European anthropologists while the harem wives (like Graffigny's Peruvian woman) provide authentic-sounding 'native scripts' (76).

girls from the North Caucasus (which is between Russia and Georgia), famed for the whiteness of their skin, and their descendants, along with many others from 'the East' as well as 'the South', continue to be sex slaves *chez nous* today.[12]

Paris

Chardin as Refugee

For Chardin, the key distinction between his experiences of hospitality in the Orient, and his experiences once he returns to Europe for good, concerns religion. While today 'we' regard the Arab world as problematically saturated with religion, and Europe (in particular the French state) as secular and therefore tolerant, for Chardin the opposite was true.[13] Chardin's religious beliefs were respected in the Muslim (Sunni and Shia), Hindu, or Orthodox Christian contexts in which he found himself. French Catholic monasteries *in the Orient* seemed to operate in a purely secular way at that time to provide hospitality to French travellers, to act as French embassies where none existed, and to show, by their very presence, the reach of the French state. They did not have a missionary function and did not seek to convert Muslims to Catholicism. A wealthy Protestant such as Chardin, with his gift for translation and knowledge of other cultures (qualities which had commercial value), was made very welcome. The France of the Sun King was a great deal less welcoming, and provides an early example of a state that turns against a whole section of its own subjects, expelling them, and thus setting in motion a significant migration of population.[14]

Cixous tells us in the programme notes to *Le dernier Caravansérail* that the term refugees (from the Latin) is first used in French and English for French Protestants

12 The history of Circassian women (once matriarchs in some tribes and proud equals in others) as objects of exchange merits detailed study. Not only, the prime consideration, were they sold into slavery in Turkey and Persia, but they also became symbolic pawns in discourse around enslavement in nineteenth-century America – where 'rescued white slaves' were often displayed in dime museums and circus side-shows.

13 'One extraordinary advantage that these peoples have over Christians is that they are not troubled by Religion [...] Each person is completely free in that respect and believes what he chooses' (Chardin II, 306; see *Lettres Persanes* XXIX, note p. 67). However, there is the issue of the schism between Sunni and Shia (which Montesquieu refers to on p. 126), and of the fate of the 'Guèbres' (Gabars) who hold to the ancient Persian religion. Elsewhere Montesquieu can be very critical of Islam, and, of course, one of his main concerns is to chastise Catholicism.

14 Louis XIV (1638) ascends throne (1643–1715) remains a key point of reference for the *Persian Letters* which cover the period 1711–1720. Persecution of Protestants does not of course end with his death as the Calas affaire brings to our attention. Many of the Persian letters have passionate diatribes against religious intolerance, and make the point that persecuting religious minorities and sending them into exile is financially disadvantageous: 'The effects of the expulsion of the Moors from Spain are as perceptible now as on the day after it occurred. Far from the gap having been filled, it is growing larger all the time' (Letter CXXI, 217/255).

(between three and 500,000 of them) fleeing religious persecution finally enshrined in the Revocation of the Edict of Nantes in 1685:

> Louis XIV auto-immunized the kingdom: on the pretext of defending religion he seized *in toto* businesses, industries, finances, arts, sciences and the rest.
>
> A primal scene for France, emblematic of so many other executions-expulsions up to the present day. The most tragic of the last century being the expulsion of the Jews under Pétain. History repeats itself, thinly disguised.

One of the Huguenots who flees to England is Chardin. During his travels in the Orient he has many opportunities to make contact with English merchants, and he nurtures contacts with the East India Company. Cixous presents England (and other Protestant nations) as suitably hospitable to Huguenot refugees unlike most nations faced with asylum-seekers today. This is true but Chardin is remarked on by the English Ambassador; he is 'the right sort' of refugee. The construction of British national identity as 'tolerant' (a slightly different version of the welcome to refugees than the French Republican one), in which the arrival of the Huguenots plays its part, is always in fact a limited tolerance. Chardin was excellent at 'fitting in'.

Chardin's life is a good example of the way in which the nation's guest will often wish to act as host to other foreigners. As a Huguenot refugee in England, warmly welcomed and indeed ennobled because of his wealth and his expertise in trade with the East, Chardin entertained many less fortunate French Protestants who sought asylum after the Revocation of the Edict of Nantes, and did not forget them even in his will. France today gives us many less happy examples of the difficulties faced by those (often '*Issus de l'immigration*') who offer hospitality to refugees if they do not have the correct documentation. The trial of Jacqueline Delthombe in 1997, for harbouring friends who did not have the correct papers, became a *cause célèbre* in France (see Gotman, 2001; Rosello, 2001). It seemed emblematic of the way in which public inhospitality (or restrictions on immigration and asylum) also enforces private inhospitality.

Today

When we look back on the Enlightenment from our 'post-colonial' position, we tend to see it as either a problem or a solution, I should like to argue that it is both. As we write or speak about hospitality in the present we have a tendency to idealize past moments or other places as a rhetorical strategy to sharpen our critique of the here and now. It is useful, I believe, to interrogate the past more closely not in order to condemn instead of praising but in order to know more about how in/hospitable mechanisms can operate. In a Persia which seemed highly welcoming to the cultured Protestant merchant Jean Chardin much of the hospitality relies on the oppression of women – not only at the explicit level such as the gang rape of dancing girls (for example II, 214–5), but implicitly. An amazingly democratic structure such as the caravanserai still relates to the harem. When respectable women cannot be seen then men tend to travel without them – unencumbered traders. England accepts Huguenot refugees from Ancien Régime France – but partly for economic reasons including commercial espionage. And both Early Modern England and Persia my not be quite as tolerant in matters of religion as they seem at first sight. In the case of England we might look at the position of Roman Catholics; in the case of (Shiite) Persia, the relation to (Sunni) Turkey or indeed the position of the Gabars, highlighted by Montesquieu in his *Persian Letters* (1960, Lettre LXVII). Revolutionary France invites those who seek asylum from unjust political regimes such as England, but its hospitality is more marked by Ancien Régime practices than they or we might care to acknowledge, including commercial espionage as well as the need to expel spies and enemies of the state. And its equality is fraternity: women are not exactly equal citizens.

Political hospitality in the present has at least two broad facets; one concerns the external frontiers of the nation state or of Europe and the question who is allowed in and under what conditions. The second concerns the more porous (but not porous enough) frontiers within the nation between those '*de souche*' and those '*issus de l'immigration*', the haves and the have-nots, men and women. Those three distinctions are all tangled together, and this brings us to the question of the banlieues or cités – the recent disturbances which have demonstrated the degree of social exclusion and alienation in France.[15] The words 'ghetto' and even 'apartheid' have been used in the press to indicate the force of the internal frontiers that operate within and around big cities. I could select three texts at random: Mehdi Charef, *Le thé au harem d'Archi Ahmed* with its untranslatable pun on Archimedes' theorem,[16] Fadela Amara's *Ni Putes Ni Soumises*, and the 2005 film *La Blessure* by Nicolas Klotz, to cover different aspects of frontiers within France and the lack of hospitality across these thresholds. The playful title of Charef's novel refers in the pun to the hospitable gesture of offering tea (one of those beverages closely tied to colonial history and to class distinction) in an imagined grand harem.[17] The disenfranchised play at an identity where power would clearly include traditional power over women – and the novel itself describes a number of situations where the socially excluded men who are the protagonists batter and socially degrade the women around them. The title also refers to the failure of the French Republican education system – the boys or young men of the banlieue do not recognize themselves in classical theorems or the universal education they are given. Amara's autobiographical/sociological study makes the point that schools should be tackling some of the questions of sexuality,

15 While there are regular incidents (such as setting cars alight) in the *banlieues*, the riots that began in October 2005 in Clichy-sous-bois, after Zyed Benna (17) and Bouna Traoré (15) were electrocuted while hiding from the police, were sufficiently large in scale to receive a significant amount of attention from politicians and from the media. More than 10,000 cars were set on fire and about 250 public buildings destroyed or seriously damaged – and of course many young people were imprisoned. The unemployment rate in Clichy is about 20 per cent, but there is no ANPE or Assédic (job centre) as a recent article in *Le Nouvel Observateur* (29 September–4 October 2006), looking at Clichy one year on, remarks (17).

16 Mercure de France, 1983.

17 See Chardin's *Du bon usage du thé*, written in 1680, for an early account of the tea trade, and how to drink tea. The noun first occurs in French dictionaries in 1789 with the wonderful innocent gloss 'they bring us the leaves' 'on nous en apporte les feuilles' (Dictionnaire de l'Académie française).

race, poverty and violence that have burning relevance in the *quartiers*.[18] In a micro-culture where love and tenderness are supposed to be hidden (a respectable girl cannot be seen holding hands with a boy) but pornographic videos and DVDs circulate, where unemployment is high, and where the macro-community withholds marks of respect or affirmation, spoken or actual violence (including sexual violence) is an obvious solution.

The advantage of the French term for both host and guest (hôte) is its reversibility – its hint in the direction of the Klossowskian guest as host of the host. Levinas's arguments about hospitality suggest that we are all guests – an ethical point similar to the philosophy behind the construction of the great caravanserais. This is one line to follow when thinking about the real make-up of the nation: how many would anyway truly be 'de souche' are not we all in some sense 'issu (e)s de l'immigration'?[19] And thinking both about those who wish to come in (one vignette brilliantly and painfully shown in *La Blessure*) and how, of those already here (all of us), some live their lives as excluded unwelcome 'guests'. The invocation of Klossowski serves to remind us, however, that the 'solutions' to inhospitality, to inequality, often involve a fraternal, indeed homosocial, bonding – we must not forget to ask questions about the place of the sister in all this.

References

Ahmed, S. (2000), *Strange Encounters* (London and New York: Routledge Books).

Alloula, M. (1981), 'The Colonial Harem' in *B. Harlow [1987]*. Godzich, M. and Godzich, W. (trans.) (Manchester: Manchester University Press Books).

Amara, F. and Zappi, S. (2004), *Ni Putes Ni Soumises* (Paris: La Découverte Books).

Asad, T. (1973), 'Two European Images of Non-European Rule' in Asad (eds).

——, eds (1973), *Anthropology and the Colonial Encounter* (New York: Humanities Press Books).

Barthes, R. (1973), *Mythologies*. trans. Lavers, A. (London: Paladin).

---

18 See Amara (2004: 52, 124–125). She argues that sexual inequality is getting worse. While her generation of women from the banlieues (militants from the 1980s) had it tough – from fathers and brothers – things seemed to be improving for them. They were not encouraged to leave home e.g. to go to university, but after much negotiation were sometimes allowed to do so (45–46). Relations with boys had to be secret, but were possible (59). She argues that lack of government investment (61) and increasing levels of social exclusion for the young men of the *banlieues*, especially those 'issus de l'immigration' has meant that 'brothers' (sometimes a designation referring to all young men of a *quartier*) terrorize their 'sisters' to a greater and increasing degree regardless whether these brothers have taken refuge in crime or fundamentalism (34ff.). The young women in their turn may adopt a range of alienated or pathological strategies (43ff.).

19 *La souche* is the tree that has been felled, and has connotations with immobility and inertia such as the expression 'sleep like a log'. It is a retrogressive form of identity, concerned with tracing back. I should say that we are all products of immigration (even if any of us think that we can trace back generations of 'pure-bloods') in the sense that the nation itself is a product of immigration and other forms of cultural mixing.

Ben Jelloun, T. (1999), *French Hospitality: Racism and North African Immigrants*, [1984 and 1987] trans. B. Bray (New York: Columbia University Press).

Chardin, J. (1671), *Le Couronnement de Soleïmaan troisième roi de Perse* (Paris: Claude Barbin).

—— (1711), *Voyages en Perse* (Amsterdam: Jean-Louis de Lorme).

—— (1811), *Voyages en Perse* (Paris : Le Normant).

—— (2002), Du bon usage du thé et des épices en Asie: Réponses à Monsieur Cabart de Villarmont, ed. I Baghdiantz Mccabe (Briare: L'inventaire).

Cixous, H. (first performed 2003), *Le dernier Caravansérail Odyssées* (The Last Caravanserai. Odysseys), Programme Notes.

Deleuze, G. and Guattari, F. (1980), *A Thousand Plateaus* [1988] trans. B. Massumi (London: Athlone).

Derrida, J. (1997), *Adieuto Emmanuel Levinas* trans. P.-A. Brault and M. Naas [1997] (Stanford: Stanford University Press).

Douthwaite, J. (1992), *Exotic Women: Literary Heroines and Cultural Strategies in Ancien Régime France* (Philadelphia: University of Pennsylvania Press).

Fassin, D., Morice, A. and Quiminal, C. (1997), *Les Lois de l'inhospitalité. Les politiques de l'immigration à l'épreuve des sans-papiers* (Paris: La Découverte Books).

Gotman, A. (2001), *Le Sens de l'hospitalité: Essai sur les fondements sociaux de l'accueil de l'autre* (Paris: PUF).

Grosrichard, A. (1979), *Structure du sérail: La Fiction du despotisme asiatique dans l'occident classique* (Paris: Seuil).

Gunny, A. (1996), *Images of Islam in Eighteenth-Century Writings* (London: Grey Seal).

Hallward, P. (2006), *Out of this World: Deleuze and the Philosophy of Creation* (London: Verso Books).

Kabbani, R. (1986), *Europe's Myths of Orient Devise and Rule* (London: Macmillan Publishing).

Kaplan, C. (1996), *Questions of Travel* (Durham and London: Duke University Press).

MacMaster, N. (1997), *Colonial Migrants and Racism. Algerians in France 1900-62* (Basingstoke: Macmillan Publishing).

Montesquieu, C. (1960), *Lettres Persanes*. Vernière, P. (ed.) (Paris: Classiques Garnier).

—— (1973), *Persian Letters*. trans. Betts, C. (Harmondsworth: Penguin Books).

Orr, M. and Sharpe, L., eds (2005), *From Goethe to Gide: Feminism Aesthetics and the Literary Canon in France and Germany 1770–1930* (Exeter: Exeter University Press).

Rosello, M. (2001), *Postcolonial Hospitality: The Immigrant as Guest* (Stanford, CA: Stanford University Press Books).

Schaub, D.J. (1995), *Erotic Liberalism Women and Revolution in Montesquieu's* 'Persian Letters' (Lanham, Maryland and London: Rowman and Littlefield).

Still, J. (2003), 'Acceptable Hospitality: from Rousseau's Levite to the Strangers in our Midst Today', *Journal of Romance Studies*, **3**(2), 1–14.

—— (2004), 'Language as Hospitality: Revisiting Intertextuality via Monolingualism of the Other', *Paragraph*, **27**(1), 113–127.

—— (2005a), 'The Stranger Within: the Figure of l'hôte in Rousseau's *Confessions*' in Orr and Sharpe (eds).

—— (2005b), 'Derrida: Guest and Host', *Paragraph*, **28**(3), 81–97.

—— (2006), 'Enlightenment Hostipitality: The Case of Chardin', *French Studies*, **40**(3), 364–368.

—— (forthcoming), 'Hospitality and Sexual Difference: from Homer to Luce Irigaray' in Tzelepis and Athanasiou (eds).

Tzelepis, E. and Athanasiou, A., eds (forthcoming), *Luce Irigaray and 'the Greeks'* (Albany, NY: SUNY Press).

Van der Cruysse, D. (1998), *Chardin Le Persan* (Paris: Fayard Books).

Wahnich, S. (1997), *L'Impossible Citoyen. L'étranger dans le discours de la Révolution française* (Paris : Albin Michel).

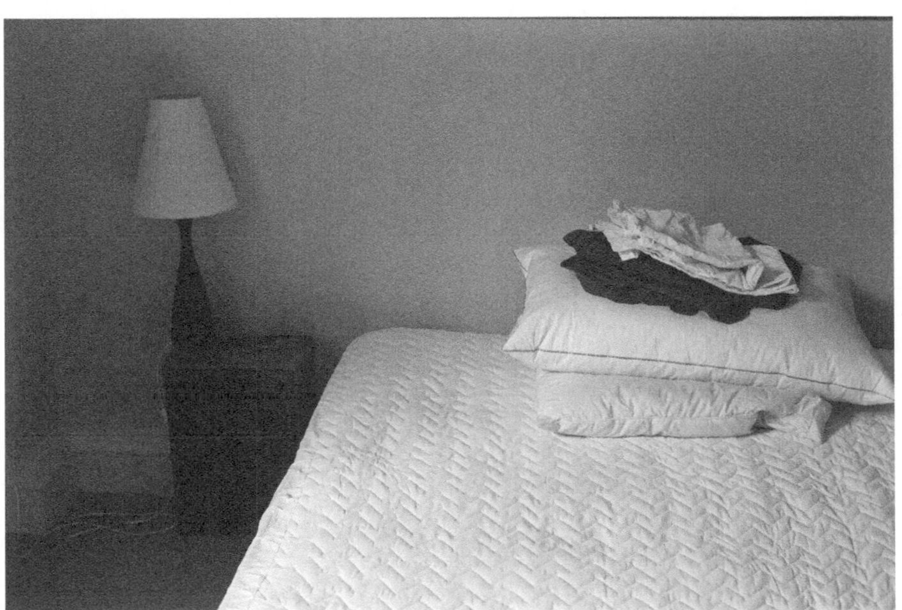

My Room Packed Up **(Hackney, UK, 2006) by Elly Clarke**

Index

abject class 170
advertising 132, 135
affect 85, 89, 109, 111
  and belonging 155–6
  see also emotion
affinity groups 86, 94, 97–9
Africa 51, 179, 181
African-Americans 47–8
Ahmed, Sara 11, 15, 77, 149, 160n1, 171
alcohol
  and drunkenness 148, 150
Amin, Ash 36
aporia
  hospitality as 5, 16, 169, 183, 188
Arab
  hospitality 20, 193
Aramberri, Julio 7, 127
Aristarkhova, Irena 74
artefacts 18, 58–9, 123, 126
  household 132–7
  see also objects
assimilation
  of immigrants in France 181, 184, 187
  of Russian immigrants in Israel 151–2, 156
asylum 4, 60, 159
  in Britain 159, 161–71, 161n3, 165n4, 166n6–7
  detention centres 1
  in Europe 186, 194
  in France 181
  and French Protestants 204–5
  and hospitality discourse 8–9
  as human right 10
  as national problem 9, 19
  see also asylum seekers
asylum seekers 2–3, 8, 10, 12–15, 19, 67, 78, 159–171, 165n5, 166n6–7, 181, 186, 194, 204
  see also refugees
Atlanta, Georgia 33
Auckland, New Zealand 34–35
Australia 13–14
authenticity
  and commercial home 124
  see also genuine hospitality

Balibar, Etienne 183–4, 184n9, 186
banlieue 202, 205, 206n18
bars 27, 35, 86, 150, 153
  as commercial hospitality space 82
  and lifestyle 94, 96
  and service culture 92
  and sociality 83, 97
  see also restaurants; Revolution Vodka Bar
Barthes, Roland 195
Bauman, Zygmunt 9, 52–3, 162n4, 187
bed & breakfast (B&B) 121, 129, 132, 135
belonging
  affective 155
  in cities 37–8
  and citizenship
  in the commercial home 121, 124
  and community 73–4
  and emotion
  and estrangement
  of guest 5
  and hospitality 18
  of host 5
  and immigrants 19, 145, 148–9, 152, 154, 157, 178
  and the Internet 3, 11
  and mobility 6, 8, 16
  national 164, 168, 183, 187
  promise of for gays 147, 157
Benedikt, Michael 40–1
Benveniste, Emile 67, 70
Blair, Tony 13, 160, 165
Blues
  electric (also known as Chicago blues) 47–8, 50, 54, 61
Blunkett, David 162, 164, 166
bodies 151, 202

of asylum seekers 14, 165–6, 168
as 'meat' 39n5
and mobility 17, 38, 112
in play 36
as recipient of hospitality 17
and senses 111
and spa treatments 17, 103–4, 110–1, 113–4
of spa workers 18, 113
see also senses
borders
and hospitality 4–5, 9, 170, 181
policing and control of 18, 165, 167, 169–70, 199
see also limits, boundaries
boundaries 3, 6, 12, 68, 74–5, 111, 133–4, 140, 148–9, 162–3, 179, 183, 187
disciplinary 2, 8
policing of 18, 66
Bourdieu, Pierre 29
Brah, Avtar 146–7
Britain 18, 185–6
eating and drinking venues in 89
and national discourses of hospitality 19, 159–71, 165n5, 166n6–7
Brown, Gordon 160

cafés 16–7, 35–8, 42, 83–7, 89, 92, 94–7
see also bars, restaurants
Caravanserai 20, 193n2, 197–200, 199n9, 203–4, 206
Caribbean 13, 32, 160
Cayman Islands 13
Chardin, Jean 19, 196–201, 196n4–7, 198n8, 202n10, 203–4, 203n13, 205n17
Chicago
Maxwell Street neighbourhood and market 47–60
Chicago City Landmarks Division 55
cities 51
and creativity 33–34
as cybercities 31
English 8, 84, 92
and migrants 48, 150, 150n7, 202, 205
as places of hospitality 16, 32, 35–6, 41, 84, 98, 199
and promotion 16, 32–3, 35
citizenship 3–4, 6, 8, 19, 164, 178–80, 186–7
Cixous, Hélène 193n2, 199, 199n9, 203–4
Clarke, Charles 165
Clifford, James 11, 123
colonialism 9, 19, 178–80, 184, 186, 189, 194
see also postcolonialism
commercial home 8, 18, 121–30, 121n1, 132–7, 139–40
commodification 104, 126
commoditization of women 201
community 35, 37, 92, 97, 99, 197, 199, 206
global 66, 73–8
and hospitality 66, 105, 127–8, 130, 136
and immigrants 145, 149, 155, 188
on the Internet 11, 17, 67–73, 154
national 163, 183
Protestant 20
queer 19, 146n3
virtual 74
computer 38–9
see also technology, Internet
concubines 201
see also commoditization of women
Conservative Party 160
consumption 17, 31, 35–7, 50, 83–7, 91, 98–9, 109, 115, 131, 150–1
conviviality 16, 35, 97
cosmopolitanism 4, 9, 17, 60, 65–6, 70, 75–7, 79, 124, 130, 179
aesthetic 76–7
and cities 16, 32–3
and mobility 2–3
Crang, Philip 87–8
creative class 33–4
creative city 34
Crouch, David 85, 88
cuisine 32, 89
see also food
customers 30, 114
and hospitality relations 7, 17, 35, 37–8, 86–91, 94, 96–9, 108, 114
cyberspace 39–40
see also technology, virtual
Certeau, Michel de 97

décor 87, 89, 96, 126, 131, 136
see also style
Deleuze, Gilles 107, 197
democracy 179, 188
Derrida, Jacques 4–6, 9, 12, 15–6, 29, 49, 53, 60, 66–7, 71n3, 73–5, 90, 104n4, 139–40, 164, 169–71, 171n11, 178,

180–3, 181n6, 182n7–8, 188, 194, 199n9
detention centres 1, 9
diaspora 1, 8, 152, 185–8
queer 147
Dikeç, Mustafa 78–9, 107
dirt 52
see also waste
discourse 33, 71, 71n3, 127–8, 134, 166, 178, 188, 195, 196n7, 199n12
of hospitality 5–11, 14, 16, 18–20, 78, 104n4, 115, 159–60, 162, 164–5, 171, 193, 199
disgust 52, 149, 152–3, 155, 161, 168, 170, 202n11
as abject
distinction
and Bourdieu, Pierre 29, 84–5
domestic
and hospitality 71n3, 123, 127, 131, 133–7
'droits de l'homme' 193
see also human rights
dwelling-in-traveling 10

economy 34, 85, 91, 152, 162, 165, 200
tourist 13
new 31
of hospitality 67, 75, 164
service 92, 95, 97

Edict of Nantes 204
emotion 31, 86, 88, 111–2, 125, 149, 155, 162, 166n7, 168, 171
enemy
guests as 66, 70, 78
host as 182
Enlightenment 181, 193, 196, 204
ethics 1–6, 8, 10, 15–6, 18–9, 66, 74, 130, 160n1, 166, 168, 170, 177–8, 180–3, 181n6, 182n8, 188, 194, 206
ethnicity 1, 14, 18, 185
excess 9, 51–3, 67, 110, 162, 165, 167, 170, 197
see also objects
exclusion 1, 9, 11, 16–7, 19, 60, 74, 77, 84, 90, 94, 108, 137, 151, 163, 166–8, 177–80, 183–7, 197, 205, 206n18
and inclusion 16–7, 84, 89, 91, 98–9, 139, 156, 186
see also inclusion

fixity 12, 54, 60, 187
flexibility 113–4, 187
flickering moments 16, 30, 37, 42
see also temporality
Florida, Richard 33–4
food 31, 35, 48, 51, 69, 77, 83–5, 87–8, 90, 92, 95, 97, 103, 108, 110, 131, 148–9, 153, 201
see also cuisine
Fortier, Anne-Marie 155
Foucault, Michel 13
France 18, 20, 161, 178, 181, 184–5, 187–8, 193–5, 202–5
and colonial past 19, 186
freeloaders 65, 67, 70
Freud, Sigmund 155–6, 156n12, 163

Galloway, Anne 40–1
garbage 48, 52–3, 58
see also waste
gender 14, 18, 29n1, 122, 128, 134, 193, 197
generosity 5, 9, 12, 19, 66–7, 70, 154, 160, 163–65, 167–71, 193, 197
geography 85
of health and leisure 112
Gilroy, Paul 185
Glasgow, Scotland 33
global
citizens 78
community 66, 73–8
globalization 1–3, 124, 194, 197
Goffman, Erving 37, 87–8
grief 155–6
see also mourning
Guattari, Felix 107, 197
guest
immigrant as 9, 19
as parasite 66–7, 70, 78, 164
performance 17, 31, 35, 37, 40, 65
see also host; host–guest relations
guest house 121–2, 127, 131, 132n2

Hage, Ghassan 15
Hannam, Kevin 15
harem 20, 198, 200, 202, 202n10–11, 204–5
see also commodification of women
Harrogate, England 17, 83, 86, 91–2, 94–5, 98–9
headscarf
ban in France 184–5, 184n9

health 91, 104–5, 109, 112, 115, 159
  and well-being 17, 103, 159
Hetherington, Kevin 13
home
  commercial 121–40
  mastery of 5, 164, 169, 178
  as mobile 11–13
  nation as 11, 19, 145–7, 151, 161, 164–6, 169–70, 171n11, 178, 181–3
homeland 7, 29, 147, 178, 185–6
homeless 1, 14, 52–3, 65, 156
Homer 194, 196, 200
homesickness 12
  see also nostalgia
hope
  and hospitality 15–6, 19, 115, 128, 154, 171, 180
hospitable space 41, 107, 112, 115
hospitableness 12, 20, 30, 32–3, 35, 38, 41, 84, 89–91, 95, 98–9, 131, 159–61, 165, 166n8, 169
hospitality
  Arab 20, 193
  absolute 4–5, 16, 75, 79, 104n4, 169–70, 181
  as aporia 5, 16, 169, 183, 188
  commercial 17, 31, 35, 37, 83–6, 91, 99, 123, 133, 135, 137, 140, 152
  conditional 4, 9, 107–8, 115–6, 169–71, 178, 183, 188
  failure of 1, 14, 18–9, 171, 184–5, 193
  and generosity 5, 9, 12, 19, 66–7, 70, 154, 160, 163–65, 167–71, 193, 197
  as gift 9, 12, 15, 67–8, 129, 154, 157, 164, 170, 200–1
  and hostility 2, 4, 9, 18–9, 140, 160–1, 163, 169, 182–3, 185, 188
  industry 7, 35, 37, 108–9, 115
  intellectual 2
  intertextual 20, 194
  the law of 2, 4–5, 170–1, 171n11, 177, 181, 183, 188
  as liminality 18, 139–40
  limits of 5, 12, 16, 18, 53, 74, 105, 164, 169–70, 177, 180
  migrant discourses of 8–10
  national 67, 78, 170
  paradox of 16, 66, 76, 123, 140, 160, 163, 169, 171, 182, 188
  political 195, 205
  postcolonial 160
  promise of 10, 19, 169, 171
  pure 107, 115, 171
  and reciprocity 4, 17, 66–9, 71, 73, 75, 127, 131, 193
  as social pact 67–8
  and technology 2, 16–7, 30, 37–42, 66, 125, 178
  as threshold 11, 18, 133, 170, 182–3, 205
  tourism discourses of 6–8
  ubiquitous 41–2
  unconditional 9, 75, 169–71, 178, 182–3, 188
  universal 2, 4, 9, 60, 179–80, 182
  virtual 38–41
host
  as mobile 30–2, 39, 66, 69, 123, 127–9
  nation as 8–9, 14, 165–6
  performance 17, 31, 35, 37, 40
  see also guest; *hôte;* host–guest relations
host–guest dichotomy 5–7, 10, 17–8, 170
host–guest relationship 5, 7–8, 10, 29, 31, 37, 41, 67, 79, 121, 123, 131–2, 135, 137, 140
hostility
  and hospitality 2, 4, 9, 18–9, 140, 160–1, 163, 169, 182–3, 185, 188
  and intolerance 160–1, 163
see also enemies; violence
hôte 5–6, 181, 206
  see also host
hotel 10, 27, 68, 111, 122, 124, 126, 130–1, 134–5, 137, 139
Howard, Michael 160
human rights 2, 4, 13–4, 167, 193
  see also 'droits de l'homme'
hybrid 39–40, 42, 124, 187, 197

immigrants
  in Britain 160, 166n9–10
  in France 184–6, 184n9
  as guest 9–10, 19, 147
  in Israel 145–57
  queer 145–52, 146n3, 152n9, 155–7
  see also migrants
immobility 3, 6, 14, 29, 38n3, 128–9, 140, 197, 206n19
inclusion 76, 108, 137, 163
  and exclusion 16–7, 84, 89, 91, 98–9, 139, 156, 186
  see also exclusion

Internet 2, 8, 11, 17, 30, 39–40, 65–7, 72–4, 132
intolerance
and Britain 160–1, 163, 166n6, 170
and Catholic France 19, 161
and hostility 160–1, 163
invitation 10, 12, 60, 65, 69, 78, 129, 178
Islam 19, 178, 184, 186, 188, 196n7, 199, 203n13
Islamophobia 185, 187
Israel 18–9, 145, 152–3, 156
Israeli 'Law of Return' 19, 145

Kant, Immanuel 4, 66, 75, 79, 139, 179–81
Kaplan, Caren 197
Klafter, David Saul 53

Labour Party 159–60
Lashley, Conrad 37, 131
Leeds, England 89, 166n7
Levinas, Emmanuel 79, 180–1, 194
lifestyle 17, 34, 85–6, 89, 94, 97–8, 125, 128–9, 132–3, 140
see also bars; cafés; restaurants; service culture
local
activists 16, 59
community 105, 124, 127–8, 130
culture 85
economy 91, 124
elections 177
history 132
hospitality 35
populations 6–7, 179
queer scene 146
residents 17, 94
roots 121
service cultures 90
venues 86, 92
see also residents
London, England 32, 52, 155, 178

Maffesoli, Michel 85–6, 88
Manchester, England 34, 84, 89–90
Massey, Doreen 12
material
artefacts 18, 58, 87
and body 39n5
culture 87–8, 99
environment 83, 86–7, 99, 109
infrastructure 14
landscape 16, 47, 50, 56–7
mobility as 49
performance of hospitality 1, 16, 68, 88
place as 47–8, 56
service cultures 88
structure 54–5, 57, 60
see also objects
McDonaldization 124, 130
melancholia 155–6
melancholy 156n12
memories 48, 88, 121, 123, 128
memory 6, 16, 47, 60, 88, 151, 185–6
Middle East 18, 51, 177
migrants 2–3, 5–7, 10, 16, 19, 33n2, 34, 47, 50, 53, 60, 160–1, 163, 165, 168, 180–3, 185–6, 189, 194, 197, 202
African-American 48
Dutch 48
German 48
Hispanic 48
Queer 18–9, 145–7
Russian 18–9, 146
migration 1–6, 8–10, 16–8, 47–8, 54, 104, 147, 165, 181, 189, 203
see also immigrants; migrants
Minority Report (movie) 40
mobile computing 16, 38
see also technology
mobile phone 37, 39, 41
see also technology
mobilities 1, 7, 10, 14, 17, 19, 29, 50, 83, 123
group 17
'micro' 17, 109, 111
migrant 6, 8, 10, 16, 18
new 16, 30, 41–2
as a paradigm 3, 30, 121n1
past 16–8, 47
and place 49–50, 52, 89
and service cultures 85, 88–9, 94, 97–8
and technology 38–9
tourist 6, 8, 10, 13, 18, 84
mobility 1–4, 6, 10–12, 14–8, 29, 31, 37, 38n3, 57, 78–9, 104–5, 108–9, 112–6, 121n1, 123–9, 132, 140, 147, 149, 159, 162, 177–8, 197–200
and bodies 104, 113
corporeal 5, 130
human 140
of objects 59
and place 16, 47–50, 52

politics of 3, 6
and technology 38–40, 66
Montesquieu 195–6, 195n3, 196n7, 202, 203n13, 205
mooring 13–5
Morley, David 11
mourning 153, 155, 156n12
see also grief
multiculturalism 3, 5, 8–10, 16, 33, 160–1, 163

narcissism 9, 14, 156, 160, 163–5
nation 1–2, 5–6, 9–12, 14, 19, 32, 49, 71, 136, 159–62, 164–6, 168–71, 178–9, 181–7, 195, 205–6
as home 11, 19, 145–7, 151, 161, 164–6, 169–70, 171n11, 178, 181–3
as host 8–9, 14, 165
nationalism 8–10, 18, 161, 163, 170, 177, 183, 185
national identity 8, 19, 159–61, 169, 185, 204
see also patriotism
National Parks Service 54–6, 58
nomad 3, 9, 20, 130, 193, 195, 197, 199–200
thought 197
non-place 124, 140
nostalgia 12, 151, 156

objects 1, 6, 12–3, 30–1, 47–51, 54, 57–9, 68, 83, 87–9, 123–4, 126–8, 136–7, 149, 154–6, 161, 162n4, 168, 170, 201, 203n12
odour 51–2, 95, 99, 111
Olympics 33n2
opening 18–9, 73–5, 78, 164, 181
and borders 5, 18, 199
and democracy 182, 188
and home 12–3, 147
see also limitations; boundaries
Orient 195–7, 196n7, 200–1, 202n10, 203–4
Orientalism 19–20, 195–7, 195n3, 201
Orthodox Christian 203

pace 109
see also rhythm; tempo
paradox 86, 123, 140
of global community 66, 73
of hospitality 16, 66, 76, 123, 140, 160, 163, 169, 171, 182, 188
parasites 10, 169
asylum seeker as 67, 168
guest as 10, 66–7, 70, 78, 164
parasitism 67
Paris, France 52, 203
patriotism 160, 182
see also national identity
performance 33, 47, 87, 136, 151–2
as guest 17, 31, 35, 37, 40, 65
of hospitality 31, 83, 94
as host 17, 31, 35, 37, 40
of lifestyle 17, 85
of service culture 87–9
Persia 196–8, 199n9, 200–1, 202n10, 203n12, 204–5
place 14, 37–8, 41, 47, 49, 50–7, 60, 97–8, 108, 127, 140, 147, 151
attachment to 121n1, 136
home as 12–13, 29, 124, 128, 130–1, 136, 146, 148
and hospitality 10–12, 15, 17, 35, 50, 60, 65, 67–8, 83–6, 88–92, 94–5, 98–9, 146–9, 152–7, 169, 179, 199
identity and 123
and memory 16, 47–8, 50, 60
and mobility 16, 47–50, 52
'of our own' 146–7, 154–6
politics 3, 10, 13, 154, 161, 170–1, 177n2, 180–1
of deterrence 159, 166
of identity 127–8, 131, 133–4
of hospitality 4–5, 178, 182–3, 182n8, 188, 194
of mobility 3, 6
pornography 202, 206
Portugal 195
postcolonial
diaspora 19, 179, 187–8
hospitality 160
migrants 182–3, 189, 193
settlers 178, 185
postcolonialism 9, 189
Powell, Enoch 160
private
home 18, 121, 123–6, 128, 135
hospitality 8, 37, 105
inhospitality 204
and public 3, 16, 18, 37–9, 126
see also public
promotion 11, 32–3, 35, 103, 114, 132
see also advertising

prostitutes 199, 201–2
    see also commoditization of women
Protestants 19–20, 160, 161n2, 196, 203–4, 203n14
proximity 11, 17, 40, 76–7, 86, 107, 149, 162, 187–8
    to strangers 6, 170
Puar, Jasbir 13
public 16, 19, 33, 35–9, 41, 49, 85, 87, 90, 92, 97, 99, 115, 124, 126, 145, 149, 154, 180, 185, 188, 193, 204
    see also private

Queer
    Queer community 19, 145–8, 151, 154–5
    queer tourism 13
    queer migrants 18, 145–8, 155–7

race 1, 14, 18, 156, 178–81, 184–5, 197, 206
racism 160, 162, 168, 177–8, 185, 193
railway 16, 30–1
    see also trains
reciprocity 4, 17, 66–70, 71, 73, 75, 127, 131, 193
refugees 2–3, 8, 10, 14, 53, 159, 161, 163–5, 167–8, 177, 193, 196–7, 199, 203–4
    see also asylum seekers
Register of Historic Places 47–8, 54–5
religion 18–9, 148, 152, 178–181, 183, 184n9, 185, 187, 189, 200–1, 203–5
reputation 17, 33, 66, 70–73, 75, 92
residents 84–5, 91, 97–9, 121, 123
    see also locals
restaurants 83–7, 95, 97, 126, 132
    see also bars, cafés
risk 66, 70, 71n3, 73, 75, 79, 160–2, 169, 170
Roche, Maurice 32
Roman Catholics 205
Rosello, Mirielle 15
Royal Baths 92, 94
    see also spas

Said, Edward 184, 196
sanatorium 109–110
Second World War 91, 189, 194
security 66, 70–3, 164, 177, 183, 185
Sellafield nuclear plant 83, 91
senses 1, 51, 88, 111–2, 114
    olfactory 52, 111
    sound 48, 51
    see also body; sensuousness
servants 41, 104, 115, 198
service 7, 30, 34, 67, 84–5, 103
    culture 17, 83– 84, 86– 92, 94, 97–99
    economy 91–92, 95
    stylized service 89–91, 94
sexuality 14, 18, 147, 151, 205
Sheller, Mimi 14–5, 29–30, 37–8, 40, 121n1, 126
Simmel, Georg 11, 140
slavery 179, 203
smell 48, 51–2, 88, 95, 111
    see also odour, olfactory senses
Smith, Valene 6
social class 37, 134
social groups 11, 17, 84, 97, 163
    see also affinity groups
social networking websites 86
social networks 2, 129
socialities 17, 83–87, 89, 91, 94–95, 97– 99, 124
sociality 17, 36–8, 40, 86, 89, 98
    empathetic 86
    network 30
soul 147–9, 152–3, 155
    Russian (dusha) 148
Souza e Silva, Adriana de 39
sovereignty 11, 180
    of host 11–2, 60
    of nation 9, 14, 162, 163, 170, 171
Soviet Union 145, 148, 149, 150, 151, 152, 154, 156
spas 8, 91, 103–6, 108–15
    see also Royal Baths
spatiality 16
    and temporality 41
stewards 31
Stockholm, Sweden 103, 110,
Stonehenge 49
stranger 1–3, 6, 8–12, 14–6, 20, 29, 36–8, 65–7, 70–3, 75, 77–9, 90, 105, 107, 109, 121, 127, 129–30, 136, 140, 147, 159–61, 164, 167, 169–70, 185, 187, 195, 197–9
    'étranger' 195
style 17, 83–6, 88–90, 94–7, 99, 109–10, 113
'style crowd' 90, 92, 94, 98–9
stylized service 89–91, 94
    see also lifestyle

Sun King 200, 203
surveillance 31, 40
  of immigrants 165, 178
  and the Internet 73
  national 33n2, 165
  and reputation systems 66, 71, 75,
  in Soviet dining centres 149
sybarites 200–203
sybaritic hospitality 20
synesthesia 110

technology 2, 16–7, 37–42, 66, 125, 178
  calculator 39
  cell phone 39
  digital still and video camera 39
  email 39
  global positioning data 39–40
    MP3 audio and video file player 39
  mobile phone 37, 39, 41
  personal organizer 39
  RFID (radio frequency identification) 40
  text messaging 39
  voice communication 39
  wi-fi 30
  wireless computing 30, 41
  and see Internet
Tel Aviv 18, 146, 148, 150–51
telepresence 39–40
  as co-presence 39
  as proximity 40
temporality 2, 29
  ephemeral 49, 89, 121
  and hospitality 16
  length of stay 130, 135
  and 'moments of hospitality' 19, 30, 32, 35, 38, 42, 66
  and permanence/permanent 10, 16, 18, 60, 121, 124, 140, 184, 200
  and spatiality 41
  'stuck in time' 150, 165
  transient/transience 59, 60, 121, 123, 137, 199
terrorism 165n5
terrorist 167
  attacks of 11 September 2001 31, 71n3, 178
  bombings in Madrid 178
  London bombings 178
Thatcher, Margaret 160
Thompson, Michael 59
threshold 11, 18, 133, 170, 182–3, 205
trains 16, 30–1
  see also railway
tolerance 5, 12, 19, 66, 106–8, 159–61, 163, 165, 168–71, 204
tourism 2–3, 6–8, 10, 13–4, 16–7, 29, 37, 83, 85, 87, 91, 99, 104, 127, 194, 197
  and hospitality 6–8
  industry 2, 7
tourists 3, 6, 7, 17, 37, 50–2, 85, 88, 121, 130, 177, 197
travel 1–2, 4, 6, 8, 11, 17–8, 29–31, 51–2, 59, 65–6, 69, 75, 78, 84, 105, 127, 136, 140, 194, 197, 204
traveller 1, 3, 7–8, 10–11, 13–4, 30–1, 49, 65–7, 69–72, 78, 121, 124–5, 130, 132, 140, 196–9, 203
traveling-in-dwelling 3, 10, 11–13
trust 70–2, 134, 178
  as condition of hospitality 70, 71
Turkey 203, 205

United States 13, 47–8, 178

urban 16, 30–1, 34, 36, 40, 52, 84–6, 92, 124, 148, 151
  hospitality 16, 20, 30, 3–4, 199–200
  lifestyle 84, 91
  space 30, 36–7, 39, 48, 85
Urry, John 14–5, 29, 30, 37–8, 49, 76, 123, 126, 162

violence 1, 3, 18, 168, 170, 180, 185, 202, 206
  and flame wars 154, 156
  and hostility 168, 170, 182
virtual 31, 40, 132
  communities 73–4
  guesting 38–41
  hospitality 17, 40
  hosting 38–41
  mobility 2, 49
  space 18, 39
Vries, Hent de 3
Wahnich, Sophie 195
walking 41
waste 1, 52–3, 60
  see also dirt; garbage; rubbish

websites 16–7, 66–73, 75–8
 see also the Internet; technology
well-being 17, 103–4, 108–14
 see also spas
Whitehaven, England 17, 83, 86, 91–2, 95–9
Wittel, Andreas 30

xenophobia 19, 177–8, 194

Yegenoglu, Meyda 5, 12

Zukin, Sharon 84–5